Lecture Notes in Mathematics

Edited by A. Dold, Heidelberg and B. Eckmann, Zürich
Series: Forschungsinstitut für Mathematik, ETH Zürich

359

Urs Stammbach
Eidgenössische Technische Hochschule, Zürich/Schweiz

Homology in Group Theory

Springer-Verlag
Berlin · Heidelberg · New York 1973

AMS Subject Classifications (1970): 20J05

ISBN 3-540-06569-5 Springer-Verlag Berlin · Heidelberg · New York
ISBN 0-387-06569-5 Springer-Verlag New York · Heidelberg · Berlin

Offsetdruck: Julius Beltz, Hemsbach/Bergstr.

INTRODUCTION

The purpose of these Notes is twofold. First, the homological algebraist may learn something about applications of homology theory to group theory, second, the group theorist may see what homological methods are able to achieve in his own field.

Chapter I introduces the reader to some basic notions in group theory. In Chapter II we have assembled the basic facts about the (co)homology of groups. Together with Chapter VI of Hilton-Stammbach [43] this will serve as a reasonable complete introduction to the (co)homology theory of groups as far as it is needed for these Notes.

Chapters III, IV, V, VI form the core of this volume. We present applications in four different but not entirely disjoint areas: extensions with abelian kernel in a variety, theorems on the lower central series, theorems on central extensions, localization of nilpotent groups.

We do not claim to be in any way complete; the choice of topics was largely determined by the preference of the author, a mild guide line being that in all four areas of applications the main homological tools are the functors $\tilde{V}$, V. These functors generalize to an arbitrary variety $\underline{\underline{V}}$ the second (co)homology group functors H^2, H_2.

It may be the place here to say something about the history of these functors. The functor H^2, H_2 make their first appearances as homology group functors in papers by Hopf [46], Eilenberg-MacLane [25], [26], Eckmann [20]. In a certain sense however they are much older; for example it is well-known that in 1904 Schur [72], [73] introduced (a group isomorphic to) the second integral homology group, the multiplicator in order to study projective representations of a group. Also, as a group of equivalence classes of extensions, $H^2(G,A)$ has been known for a long time to group theorists (Schreier [71], Baer [5], a.o.). The functors $\tilde{V}$, V for the variety $\underline{\underline{V}} = \underline{\underline{Ab}}$ have been known for a long time, also, for they coincide basically with the functors Ext^1_Z , Tor^Z_1 and as such are fundamental in homological algebra. Attempts to define functors $\tilde{V}$, V for arbitrary varieties $\underline{\underline{V}}$ are plentiful. It was clear from the beginning that the equivalence classes of those extensions of a given group by an abelian kernel, that have the property to lie in $\underline{\underline{V}}$, form an abelian group under Baer addition. More sophisticated definitions for $\tilde{V}$, V have been given by Gerstenhaber [32], Knopfmacher [50], [51]. In particular, the paper [50] already contains many of the fundamental properties of the functor $\tilde{V}$ associated with a variety $\underline{\underline{V}}$. A definition for a homology theory that works in quite general circumstances (a category furnished with a triple) has later been given by Beck [16]. Subsequently various authors have succeeded in defining similar theories: André [1], Bachmann [4], Barr [8], Barr-Beck [9], Rinehart [68], Ulmer [83]. All these categorical theories not only yield functors $\tilde{V}$, V but also functors that correspond to the higher dimensional (co)homology groups. However, none of these authors makes explicit references to varieties of groups. In Stammbach [78] a new definition of V for a variety $\underline{\underline{V}}$ of groups appears and applications to the lower central series are given. Later, in Leedham-Green [53], [54], [55] and in Leedham-Green-Hurley [56] the categorical theories in the case of varieties of groups are discussed in much detail.

Our approach is in the spirit of Stammbach [78], [79]. We are mainly interested in presenting applications of homological ideas and therefore have tried to be as elementary as possible. In particular, we have avoided, where possible, the use of high-powered categorical and homological tools, such as for example spectral sequences. To achieve this goal we clearly had to make certain sacrifices, the most important probably being that we have had to restrict our considerations to $\tilde{V}$, V , leaving aside completely the higher dimensional (co)homology groups.

We would like to mention two texts on (co)homology of groups that have objectives similar to those of our Notes, namely, that of presenting applications of homological results in pure group theory.

[35] Gruenberg, K.: Cohomological Topics in Group Theory, Lecture Notes in Mathematics, vol.143. Springer 1970.

[3] Babakhanian, A.: Cohomological Methods in Group Theory, Marcel Dekker 1972.

Although the intersection of the material covered in these three texts is not trivial, it has been kept to a minimum; either the applications presented in our Notes are in an area not covered in [35] or [3], or else the presentation is different.

Within a given chapter we use two series of enumerations, one for theorems, lemmas, propositions and corollaries, the other for displayed formulas. The system of enumerations in each of these series consists of a pair of numbers, the first refering to the section, the second to the particular item. Thus in Section 2 of Chapter V we have Proposition 2.4 in which a displayed formula appears which is labelled (2.10). If we wish to refer to a theorem, etc. or a displayed formula, we simply use the same system of enumeration provided the item to be cited occurs in the same chapter. If it occurs in a different chapter we will preceed the two numbers by a Roman numeral specifying the chapter.

It is a pleasure to make many acknowledgements. First of all I would like to express my gratitude to my teachers and friends Beno Eckmann,

Karl Gruenberg, Peter Hilton, without whom the writing of this volume would not have been possible. Very special thanks are due to Peter Hilton who read the whole manuscript; his expert advice in the area of mathematics as well as in the area of linguistics led to numerous improvements.

I also thank the editors of Springer Lecture Notes for accepting the text in their series. Finally, my thanks are due to Frau Eva Minzloff for having done such an excellent job in converting my illegible manuscript into a neat typescript.

Eidgenössische Technische Hochschule

8006 Zürich

Juli 1973

TABLE OF CONTENTS

CHAPTER I

VARIETIES OF GROUPS

In this chapter we assemble some definitions in group theory that are used in the later parts of these notes. In particular we introduce the basic notions about varieties of groups. We use, whenever possible, the language of category theory. Most of the facts contained in this chapter are well known to those even vaguely familiar with group theory. However the functorial point of view adopted here will possibly be unfamiliar to some. It will become clear in the subsequent chapters that the categorical approach has great advantages.

I.1. Some Definitions in Group Theory

We denote by **Gr** the category of groups. It is well known that **Gr** has free objects. A group F is called <u>free</u> on the set $S \subset F$ if, to every group G and to every function $f : S \to G$, there exists a unique homomorphism $f' : F \to G$ extending f.

(1.1)
$$\begin{array}{ccc} S & \xrightarrow{\;\; f \;\;} & G \\ \cap & \nearrow_{f'} & \\ F & & \end{array}$$

Given the set S this <u>universal property</u> (1.1) characterizes the free group on S up to isomorphism. The actual construction of free groups shows that if $S = (x_i)$, $i \in I$, then the elements of F may be represented as words, i.e. finite sequences in x_i and x_i^{-1}, $i \in I$.

Every group G is isomorphic to a quotient of some free group F; $G \cong F/R$. The corresponding short exact sequence of groups $R \rightarrowtail F \twoheadrightarrow G$

is called a (free) presentation of G .

Given G and a nonnegative integer q we may define the lower central (q) series $\{G_n^q\}$ as follows.

$$(1.2)\qquad G_1^q = G \ , \quad G_{i+1}^q = G \#_q G_i^q \ , \quad i = 1,2,\ldots .$$

where we use for any subgroup $N \subseteq G$ the symbol $G \#_q N$ to denote the subgroup of G generated by

$$xyx^{-1}y^{-1}z^q \ , \quad x \in G \ , \quad y,z \in N.$$

Note that if N is normal in G , so is $G \#_q N$. Note also that for $q = 0$ we simply get the lower central series

$$(1.3)\qquad G_1^o = G \ , \quad G_{i+1}^o = [G,G_i^o] \ , \quad i = 1,2,\ldots$$

We call a group G nilpotent (q) if there exists $n \geq 1$ with $G_n^q = e$. The class c of a nilpotent (q) group is characterized by $G_c^q \neq e$, $G_{c+1}^q = e$. Note that nilpotent (O) is just nilpotent.

LEMMA 1.1. *Let $q > 0$. The group G is nilpotent (q) if and only if it is nilpotent and of finite q exponent.*

PROOF. Let G be nilpotent (q) of class c . Then clearly G is nilpotent and every element $x \in G$ has exponent dividing q^c . Conversely, let G be nilpotent of class c_1 , say, and of finite q exponent q^{c_2} , say. Set $c = c_1 \cdot c_2$. It is then clear that $G_{c+1}^q = e$.

The upper central series $\{Z_nG\}$ is defined as follows

(1.4) $$Z_oG = e \ , \ Z_1G = ZG \ , \ Z_nG/Z_{n-1}G = Z(G/Z_{n-1}G)$$

where ZG denotes the center of G . If G is nilpotent of class c then $Z_{c-1}G \neq G$, $Z_cG = G$, and conversely.

The derived series $\{G^n\}$ of G is defined by

(1.5) $$G^1 = G \ , \ G^n = [G^{n-1},G^{n-1}] \ .$$

A group is called soluble if there exists n with $G^n = e$. The soluble length ℓ of G is characterized by $G^\ell \neq e$, $G^{\ell+1} = e$.

A generalization of both the lower central series and the derived series is as follows. Let $(c_1,c_2,\ldots,c_n)$ be a sequence of integers $c_i \geq 1$. We define a normal subgroup $G_{(c_1+1,c_2+1,\ldots,c_n+1)}$ of G by setting for $1 \leq k \leq n$

(1.6) $$G_{(c_1+1,c_2+1,\ldots,c_k+1)} = (G_{(c_1+1,c_2+1,\ldots,c_{k-1}+1)})_{c_k+1} \ .$$

If $G_{(c_1+1,\ldots,c_n+1)} = e$, then G is called polynilpotent of class row $\leq (c_1,\ldots,c_n)$. Note that polynilpotent of class row $\leq (c)$ is just nilpotent of class $\leq c$, and that polynilpotent of class row $\leq (2,2,\ldots,2)$, ℓ-times, is just soluble of length $\leq \ell$.

Let P be a property of groups (nilpotent, finite, etc.). Then a group G is called residually P if for every $e \neq x \in G$ there exists a normal subgroup $N \triangleleft G$ with $x \notin N$ and G/N having property P .

Given the group G and the (left) G-module A we shall denote their semi-direct product by $A \rtimes G$. The elements of $A \rtimes G$ are pairs

(a,x) , $x \in G$, $a \in A$ and the product is given by

(1.7) $$(a,x)\cdot(a',x') = (a+xa',xx') , \quad a,a' \in A , \; x,x' \in G .$$

The <u>direct product</u> of two groups G_1,G_2 is denoted by $G_1 \times G_2$, their <u>free product</u> by $G_1 * G_2$.

I.2. Definition of a Variety

A <u>variety</u> $\underline{\underline{V}}$ of groups is a full subcategory of $\underline{\underline{Gr}}$ which is closed under taking subobjects, quotient objects and arbitrary categorical products (called cartesian products in [64]).

A <u>law</u> in n-variables $v = v(x_1,x_2,\dots,x_n)$ is a finite sequence of letters $x_1,\dots,x_n,x_1^{-1},\dots,x_n^{-1}$. Clearly a law determines an element $[v]$ in $F_\infty = F(x_1,\dots,x_n,\dots)$, the free group on $x_1,x_2,\dots,x_n,\dots$. A law v is said to <u>hold</u> in a group G if for every homomorphism $f : F_\infty \to G$, $f[v] = e$.

Given a set (v) of laws, then clearly the groups G in which all the laws $v \in (v)$ hold form a variety. It is a well-known theorem of Birkhoff (see [64]) that the converse is true, also.

<u>THEOREM 2.1</u>. <u>Every variety</u> $\underline{\underline{V}}$ <u>can be described by a set of laws</u>.

We continue with a series of examples of varieties.

(i) $\underline{\underline{V}} = \underline{\underline{Ab}}$, the category (variety) of <u>abelian groups</u>. The corresponding law is $[x_1,x_2]$.

(ii) $\underline{\underline{V}} = \underline{\underline{N}}_c$, the variety of <u>nilpotent groups</u> of class $\leq c$. The corresponding law is $[x_1,[x_2,[x_3,\dots[x_c,x_{c+1}]]]]$.

(iii) $\underline{\underline{V}} = \underline{\underline{S}}_\ell$, the variety of <u>soluble groups</u> of length $\leq \ell$.

(iv) $\underline{\underline{V}} = \underline{\underline{P}}_{(c_1,\ldots,c_n)}$, $c_i \geq 1$, the variety of <u>polynilpotent groups</u> of class row $\leq (c_1,\ldots,c_n)$.

(v) $\underline{\underline{V}} = \underline{\underline{B}}_q$, the variety of groups of exponent q . The corresponding law is x_1^q .

(vi) It is obvious that the intersection of two varieties is again a variety. In particular we have $\underline{\underline{V}} = \underline{\underline{Ab}} \cap \underline{\underline{B}}_q = \underline{\underline{Ab}}_q$, the variety of abelian groups of exponent q . The corresponding law is $[x_1,x_2]x_3^q$.

A variety $\underline{\underline{V}}$ is said to be of <u>exponent</u> $q \geq 0$ if and only if $\underline{\underline{Ab}} \cap \underline{\underline{V}} = \underline{\underline{Ab}}_q$. In particular, $\underline{\underline{V}}$ is of exponent zero if and only if it contains $\underline{\underline{Ab}}$.

Let $v = v(x_1,\ldots,x_n)$ be a law and let $a_1,a_2,\ldots,a_n$ be elements in G . Then we may consider the element

(2.1) $$a = v(a_1,a_2,\ldots,a_n) \in G .$$

We say that a is obtained by evaluating v at $(a_1,a_2,\ldots,a_n)$. Note that a is the image of $[v] \in F_\infty$ under a homomorphism $f : F_\infty \to G$ defined by $f(x_i) = a_i$, $i = 1,\ldots,n$, and $f(x_j)$ arbitrary for $j > n$.

Let (v) be a set of laws defining $\underline{\underline{V}}$ and let G be an arbitrary group. We define the <u>verbal subgroup</u> VG of G associated with $\underline{\underline{V}}$ as follows. It is the subgroup of G generated by all elements of the form

(2.2) $$v(a_1,\ldots,a_n) , \quad v \in (v) , \quad a_1,\ldots,a_n \in G .$$

Note that n depends on v . We shall see below that VG only depends on $\underline{\underline{V}}$ and not on the choice of laws defining $\underline{\underline{V}}$. If $g : G \to G$ is an endomorphism, we have

$$g(v(a_1,a_2,\ldots,a_n)) = v(ga_1,ga_2,\ldots,ga_n) \ . \tag{2.3}$$

Hence the group VG is invariant, in particular it is normal. The quotient group G/VG is clearly in $\mathbf{V}$; indeed it is the largest quotient of G lying in $\mathbf{V}$, more precisely:

Every group homomorphism $f : G \to Q$ with Q in $\mathbf{V}$ factors uniquely through $G \to G/VG$

$$\begin{array}{ccc} G & \xrightarrow{\quad f \quad} & Q \\ \downarrow & \nearrow_{f'} & \\ G/VG & & \end{array} \tag{2.4}$$

We may rephrase this fact in the language of category theory as follows. The functor $P : \mathbf{Gr} \to \mathbf{V}$ defined by

$$PG = G/VG \tag{2.5}$$

is left adjoint to the embedding functor $E : \mathbf{V} \to \mathbf{Gr}$. It is apparent from this property that G/VG and hence VG depends only on $\mathbf{V}$ and not on the chosen set of laws defining $\mathbf{V}$.

We conclude this section with a couple of examples.

(i) Let $\mathbf{V} = \mathbf{Ab}$. Then the functor $P : \mathbf{Gr} \to \mathbf{Ab}$ is just the abelianizing functor $PG = G_{ab}$, and $VG = [G,G]$.

(ii) Let $\mathbf{V} = \mathbf{N}_c$. Then $VG = G_{c+1}$, and $G/VG = G/G_{c+1}$ is the largest quotient of G that is nilpotent of class $\leq c$.

I.3. Free Groups in a Variety; the Coproduct in a Variety

A group F in $\mathbf{V}$ is called $\mathbf{V}$-*free* on the set $S \subset F$ if it satisfies the following universal property: To every group G in $\mathbf{V}$ and to every function $f : S \to G$ there exists a unique homomorphism

$f' : F \to G$ extending f

(3.1)
$$\begin{array}{ccc} S & \xrightarrow{\quad f \quad} & G \\ \cap & \nearrow_{f'} & \\ F & & \end{array}$$

Given S, the above property characterizes the $\underline{\underline{V}}$-free group on S up to isomorphism. In order to construct $F = F(S)$ let $\tilde{F} = \tilde{F}(S)$ be the (absolutely) free group on the set S. Then $F = \tilde{F}/V\tilde{F}$ has the required universal property; for let $f : S \to G$ be a function, then there exists a unique $\tilde{f}' : \tilde{F} \to G$ and hence a unique $f' : F \to G$ extending f

(3.2)
$$\begin{array}{ccc} S & \xrightarrow{\quad f \quad} & G \\ \cap & & \\ \tilde{F} & \xrightarrow{\tilde{f}'} & G \\ \downarrow & \nearrow_{f'} & \\ F=\tilde{F}/V\tilde{F} & & \end{array}$$

Of course, $\underline{\underline{Gr}}$-free groups are just free, $\underline{\underline{Ab}}$-free groups are just free abelian, $\underline{\underline{N}}_c$-free groups are just free nilpotent of class c, etc. It is clear that every group G in $\underline{\underline{V}}$ may be represented as a quotient of a $\underline{\underline{V}}$-free group F, i.e. $G \cong F/R$. The associated short exact sequence

(3.3)
$$R \rightarrowtail F \twoheadrightarrow G$$

is called a $\underline{\underline{V}}$-*free presentation* of G.

Later on we shall need the following result on $\underline{\underline{V}}$-free groups.

PROPOSITION 3.1. A group F *is* $\underline{\underline{V}}$-*free on the set* $S \subset F$ *if and only if* F *is generated by* S *and, for every finite subset* T *of* S, *the subgroup* U *generated by* T *is* $\underline{\underline{V}}$-*free on* T.

PROOF: If F is $\underline{\underline{V}}$-free on S, then clearly F is generated by S and the subgroup U generated by $T \subset S$ is $\underline{\underline{V}}$-free on T. To prove the converse let G be any group in $\underline{\underline{V}}$ and let $f : S \to G$ be a function. To define $f' : F \to G$ consider $x \in F$. Clearly x is contained in a subgroup U generated by a finite subset T of S. The restriction $f_T : T \to G$ of f to T gives rise to a unique $f_T' : U \to G$. Define $f'(x) = f_T'(x)$. We have to show that $f'(x)$ is well-defined. Thus let $x \in U_i$, $i = 1,2$, where U_i is generated by the finite subset T_i of S. Then $x \in V$ where V is generated by $T_1 \cup T_2$. Since the extensions f'_{T_1}, f'_{T_2}, $f'_{T_1 \cup T_2}$ are uniquely determined we have

$$f'_{T_1}(x) = f'_{T_1 \cup T_2}(x) = f'_{T_2}(x)$$

so that f' is indeed well-defined. Clearly f' is a homomorphism. Finally, since S generates F the uniqueness of f' is obvious.

Finally we show that a variety $\underline{\underline{V}}$ always has coproducts. We restrict ourselves to the case of a coproduct of two objects; it is clear how to extend the definition to the general case. Let G_i, $i = 1,2$ be groups in $\underline{\underline{V}}$. Their coproduct in $\underline{\underline{V}}$, (also called verbal or varietal product) $G_1 *_V G_2$ is given by

$$G_1 *_V G_2 = G_1 * G_2/V(G_1 * G_2) \tag{3.4}$$

where $G_1 * G_2$ denotes the free product of G_1 and G_2. The injections $j_i : G_i \to G_1 *_V G_2$ are given by the composition

$$j_i : G_i \to G_1 * G_2 \to G_1 *_V G_2 .$$

We show that $G_1 *_V G_2$ satisfies the required universal property. Thus let Q be a group in $\underline{\underline{V}}$ and let $f_i : G_i \to Q$ be homomorphisms.

Then there exists a unique $f' : G_1 * G_2 \to Q$ and hence a unique $f : G_1 *_V G_2 \to Q$ such that the diagram

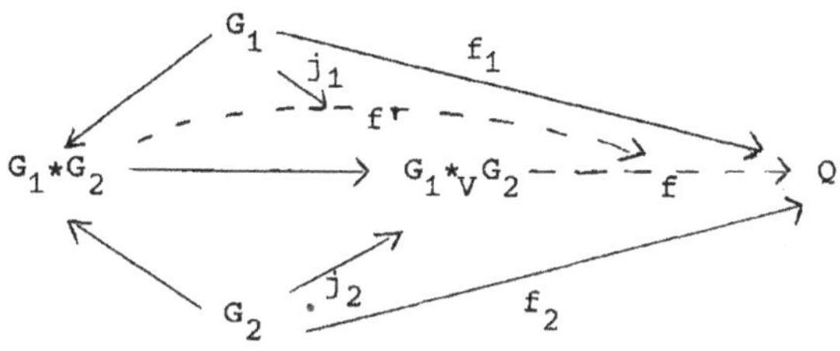

is commutative.

We finally remark that j_i has a left inverse; the left inverse of j_1 is obtained by

(3.5)

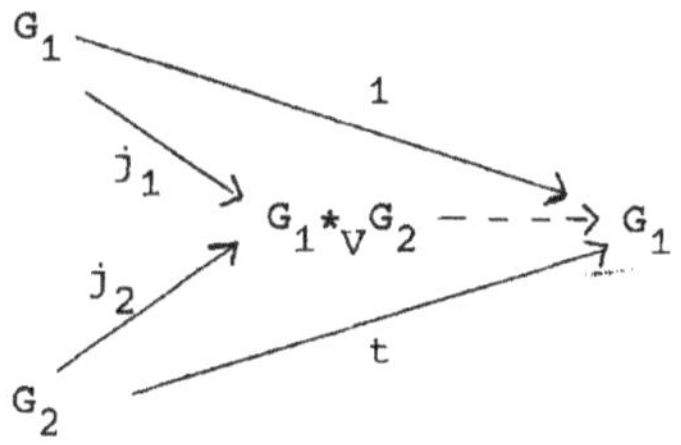

where t is the trivial map. Analogously, the left inverse of j_2 is obtained. It follows in particular that j_i , $i = 1,2$ is injective.

CHAPTER II

ELEMENTS OF HOMOLOGY THEORY

Our objective in this chapter is twofold. First we want to introduce some notation and assemble some basic facts about the homology theory of groups. We present these facts without proofs, refering whenever possible to Hilton-Stammbach [43], Chapter VI. Our second objective is to state some results about the homology of groups that are classical, but for some reason are not covered in [43] or in most of the other introductory texts. Of course, for those results we give complete proofs.

II.1. Definition of (Co)Homology

Let G be a group written multiplicatively. We denote by $\mathbb{Z}G$ its integral group ring with augmentation $\varepsilon : \mathbb{Z}G \to \mathbb{Z}$ and augmentation ideal $IG = \ker \varepsilon$.

We denote by $\underline{\underline{Mod}}_G^{\ell}$ the category of left and by $\underline{\underline{Mod}}_G^{r}$ the category of right G-modules. Let A be a left and let B be a right G-module. Then, for every $n \geq 0$, the n-th cohomology group with coefficients in A is defined by

(1.1) $$H^n(G,A) = \mathrm{Ext}^n_{\mathbb{Z}G}(\mathbb{Z},A) ,$$

and the n-th homology group of G with coefficients in B is defined by

(1.2) $$H_n(G,B) = \mathrm{Tor}_n^{\mathbb{Z}G}(B,\mathbb{Z}).$$

In both formulas, $\mathbb{Z}$, the additive group of the integers is to be regarded as a trivial left G-module.

It is clear from (1.1) and (1.2) that any short exact sequence of G-modules gives rise to a long exact sequence in (co)homology.

We remark (correcting a series of misprints in [43], p.190) that $H^n(-,-)$ may be regarded as a (contravariant) functor on the category $\underline{\underline{G}}^*$. The objects of $\underline{\underline{G}}^*$ are pairs (G,A) with G a group and A a G-module. The morphisms

$$(f,\alpha) : (G,A) \to (G',A')$$

in $\underline{\underline{G}}^*$ consist of a group homomorphism $f : G \to G'$ and a homomorphism $\alpha : A' \to A$ of G-modules. Here A' is to be regarded as a G-module via f. The induced homomorphism

$$(f,\alpha)^* : H^n(G',A') \to H^n(G,A)$$

is given by the composition

$$H^n(G',A') \xrightarrow{\theta=f^*} H^n(G,A') \xrightarrow{\alpha_*} H^n(G,A)$$

where $\theta = f^*$ is the obvious "change-of-rings" map.

Analogously we may define a category $\underline{\underline{G}}_*$ such that $H_n(-,-)$ may be regarded as a (covariant) functor on $\underline{\underline{G}}_*$. We leave the details to the reader.

II.2. Derivations

A function $d : Q \to A$ where A is a Q-module is called a <u>derivation</u> if

$$(2.1) \qquad d(x \cdot y) = dx + x dy \quad , \quad x,y \in Q .$$

The set of all derivations from Q to A obviously has a natural abelian group structure. Denoting this group by $Der(Q,A)$ we may define a functor

$$\mathrm{Der}(Q,-) : \underline{\underline{\mathrm{Mod}}}_Q^{\ell} \to \underline{\underline{\mathrm{Ab}}} .$$

By [43], Theorem VI.5.1 we have:

The functor $\mathrm{Der}(Q,-)$ is representable; more precisely there is a natural equivalence η of functors with

$$(2.2) \qquad \eta_A : \mathrm{Der}(Q,A) \tilde{\to} \mathrm{Hom}_Q(IQ,A) ,$$

given by $\eta_A(d)(x-1) = dx$, $x \in Q$, $d \in \mathrm{Der}(Q,A)$.

For fixed A we may define a functor in the obvious way

$$\mathrm{Der}(-,A) : \underline{\underline{\mathrm{Gr}}}/Q \to \underline{\underline{\mathrm{Ab}}}$$

where $\underline{\underline{\mathrm{Gr}}}/Q$ denotes the category of groups over Q . For $g : G \to Q$ the module A is to be regarded as a G-module via g . It follows from Corollary VI.5.4 in [43] that this functor is corepresentable; more precisely, there is a natural equivalence μ of functors with

$$(2.3) \qquad \mu_G : \mathrm{Der}(G,A) \tilde{\to} \mathrm{Hom}_{\underline{\underline{\mathrm{Gr}}}/Q}\left(\begin{matrix} G \\ \downarrow \\ Q \end{matrix} , \begin{matrix} A \rtimes Q \\ \downarrow \\ Q \end{matrix} \right)$$

given by

$$(2.4) \qquad (\mu_G(d))(y) = (dy,gy) , \ y \in G , \ d \in \mathrm{Der}(G,A) .$$

Here $A \rtimes Q$ denotes the semi-direct product of Q by A (see Section I.1). As an application we reprove

<u>PROPOSITION 2.1</u>. <u>Let</u> F <u>be a free group on</u> $S = (x_i)$, $i \in I$. <u>Then</u> IF <u>is a free</u> F-<u>module on</u> $S-1 = (x_i-1)$, $i \in I$.

<u>PROOF</u>: Given an F-module A and a function $f : (S-1) \to A$ we must show that there exists a unique homomorphism $f : IF \to A$ of F-modules extending f . We first note that S-1 generates IF as F-module, so that uniqueness is clear. To prove existence consider the function $h : S \to A \wr F$ given by

$$h(s) = (f(s-1),s) \quad , \quad s \in S .$$

The fact that F is free yields a group homomorphism $h' : F \to A \wr F$. Clearly, h' composed with the projection $A \wr F \to F$ is the identity on F , so that h' may be regarded as a morphism in $\underline{\underline{Gr}}/F$. By (2.3), (2.4) it gives rise to a derivation $d : F \to A$ with

$$d(s) = f(s-1) \quad , \quad s \in S .$$

The derivation d in turn corresponds by (2.2) to a module homomorphism $\varphi : IF \to A$ with

$$\varphi(s-1) = d(s) = f(s-1) \quad , \quad s \in S .$$

Thus φ is the required extension of f .

In calculations in the (co)homology of groups the so called <u>Fox derivatives</u> or <u>partial derivatives</u> are often useful (see [29]). They arise as follows. Let F be free on the set $S = (x_i)$, $i \in I$. Then by (2.2)

$$\eta_{ZF} : Der(F,ZF) \tilde{\to} Hom_F(IF,ZF) .$$

By Proposition 2.1 the module IF is F-free on (S-1) . The i-th Fox-derivative is defined to be the derivation $\partial_i : F \to ZF$ corresponding to the i-th projection

$$IF \cong \bigoplus_{i\in I} (ZF)_i \xrightarrow{\pi_i} ZF .$$

We conclude that if

(2.5) $$(x-1) = \sum_{i\in I} \alpha_i (x_i - 1) \quad , \quad x \in F$$

then

(2.6) $$\partial_i(x) = \alpha_i \quad , \quad i \in I .$$

We may express this result in the following way.

PROPOSITION 2.2. $(x-1) = \sum_{i\in I} \partial_i(x)(x_i - 1) \quad , \quad x \in F .$

Note also, that

(2.7) $$\partial_i(x_j) = \delta_{ij} e \ , \quad i,j \in I .$$

II.3. The 5-Term Sequences

Let N be a normal subgroup in G with quotient group Q. We summarize this information in the short exact sequence of groups

(3.1) $$E : N \overset{h}{\rightarrowtail} G \overset{g}{\twoheadrightarrow} Q .$$

Let A denote a left Q-module and let B denote a right Q-module. Then the following sequences are exact (see [43], p.202):

(3.2) $$0 \to Der(Q,A) \xrightarrow{g^*} Der(G,A) \xrightarrow{h^*} Hom_Q(N_{ab},A) \xrightarrow{\delta^*_E} H^2(Q,A) \xrightarrow{g^*} H^2(G,A) \quad ,$$

(3.3) $$0 \to H^1(Q,A) \xrightarrow{g^*} H^1(G,A) \xrightarrow{h^*} Hom_Q(N_{ab},A) \xrightarrow{\delta^*_E} H^2(Q,A) \xrightarrow{g^*} H^2(G,A) \quad ,$$

(3.4) $$H_2(G,B) \xrightarrow{g_*} H_2(Q,B) \xrightarrow{\delta^E_*} B \otimes_Q N_{ab} \xrightarrow{h_*} B \otimes_G IG \xrightarrow{g_*} B \otimes_Q IQ \to 0 \ ,$$

(3.5) $$H_2(G,B) \xrightarrow{g_*} H_2(Q,B) \xrightarrow{\delta^E_*} B \otimes_Q N_{ab} \xrightarrow{h_*} H^1(G,B) \xrightarrow{g_*} H^1(Q,B) \to 0 \ .$$

Here $N_{ab} = N/[N\ N]$ is to be regarded a left Q-module via conjugation, i.e.

(3.6) $$y(u[N,N]) = xux^{-1}[N,N]$$

where $u \in N$ and $x \in G$ represents $y \in Q$.

We note the important fact that all of these sequences are natural. In order to make this statement more precise let

$$\begin{array}{lccccc} E' : & N' & \rightarrowtail & G' & \twoheadrightarrow & Q' \\ & f_1\downarrow & & f_2\downarrow & & f_3\downarrow \\ E : & N & \rightarrowtail & G & \twoheadrightarrow & Q \end{array}$$

be a map of extensions. Moreover let $\alpha : A \to A'$ be a Q-module homomorphism. Then the following diagram is commutative.

(3.7)
$$\begin{array}{ccccccccccc} 0 & \to & H^1(Q,A) & \to & H^1(G,A) & \to & \mathrm{Hom}_Q(N_{ab},A) & \overset{\delta^*_E}{\to} & H^2(Q,A) & \to & H^2(G,A) \\ & & \alpha_*\downarrow & & \alpha_*\downarrow & & \alpha_*\downarrow & & \alpha_*\downarrow & & \alpha_*\downarrow \\ 0 & \to & H^1(Q,A') & \to & H^1(G,A') & \to & \mathrm{Hom}_Q(N_{ab},A') & \overset{\delta^*_E}{\to} & H^2(Q,A') & \to & H^2(G,A') \\ & & f_3^*\downarrow & & f_2^*\downarrow & & f_1^*\downarrow & & f_3^*\downarrow & & f_2^*\downarrow \\ 0 & \to & H^1(Q',A') & \to & H^1(G',A') & \to & \mathrm{Hom}_{Q'}(N'_{ab},A') & \overset{\delta^*_{E'}}{\to} & H^2(Q',A') & \to & H^2(G',A') \end{array}$$

A similar statement is true for the other 5-term sequences (3.2), (3.4), (3.5); we leave the details to the reader.

In these Notes the homology with coefficients in $B = Z/qZ$ with q any non-negative integer will play a central role. We shall therefore use the following simplified notation

(3.8) $$H_nG = H_n(G,Z) \quad , \quad n = 0,1,\dots ;$$

(3.9) $$H_n^q G = H_n(G, \mathbb{Z}/q\mathbb{Z}) \quad , \quad n = 0,1,\dots \; , \; q = 0,1,\dots \; .$$

Using the notation introduced in Section I.1 we have

(3.10) $$\mathbb{Z}/q\mathbb{Z} \otimes_Q N_{ab} = N/G \#_q N \; ,$$

(3.11) $$H_1^q G = \mathbb{Z}/q\mathbb{Z} \otimes G_{ab} = G/G \#_q G \; .$$

The latter may even be abbreviated to

(3.12) $$H_1^q G = G_{ab}^q \; .$$

With these notational conventions the exact sequence (3.2) (or (3.3)) with $B = \mathbb{Z}/q\mathbb{Z}$, $q = 0,1,2,\dots$ reads

(3.13) $$H_2^q G \xrightarrow{g_*} H_2^q Q \xrightarrow{\delta_*^E} N/G \#_q N \xrightarrow{h_*} G_{ab}^q \xrightarrow{g_*} Q_{ab}^q \longrightarrow 0 \; .$$

When $q = 0$, the superscript q may be dropped, and we may write $[G,N]$ for $G \#_o N$.

We finally recall Hopf's formula. Let $R \rightarrowtail F \twoheadrightarrow Q$ be a free presentation of Q , then sequence (3.13) reads

$$0 \to H_2^q Q \to R/F \#_q R \to F_{ab}^q \to Q_{ab}^q \to 0$$

so that we obtain

(3.14) $$H_2^q Q \cong F \#_q F \cap R/F \#_q R \; .$$

For $q = 0$ this formula is due to Hopf [45].

II.4. Extensions with Abelian Kernel

Let $N \rightarrowtail G \twoheadrightarrow Q$ be a short exact sequence of groups. Then N_{ab} is given a Q-module structure by (3.6). Let A be a Q-module. The short exact sequence

(4.1) $$E : A \overset{h}{\rightarrowtail} G \overset{g}{\twoheadrightarrow} Q$$

is called an extension of Q by the Q-module A if the Q-module structure of A defined by (3.6) agrees with the one already given in A . The extension E is called equivalent to the extension $E' : A \rightarrowtail G' \twoheadrightarrow Q$ if there exists $f : G \to G'$ such that the diagram

(4.2) $$\begin{array}{ccccc} A & \rightarrowtail & G & \twoheadrightarrow & Q \\ \| & & \downarrow f & & \| \\ A & \rightarrowtail & G' & \twoheadrightarrow & Q \end{array}$$

is commutative.

It is well-known that $H^2(Q,A)$ classifies extensions of Q by A , i.e. the set of equivalence classes of extensions of Q by A is in one-to-one correspondence with the set underlying $H^2(Q,A)$. The map Δ establishing this correspondence may be described as follows (see [43], p.207). In the 5-term sequence in cohomology associated with the extension (4.1) and the Q-module A we have the homomorphism

(4.3) $$\delta_E^* : \operatorname{Hom}_Q(A,A) \to H^2(Q,A) .$$

Then the cohomology class $\Delta[E] = \xi \in H^2(Q,A)$ associated with the equivalence class [E] of the extension E is given by

(4.4) $$\Delta[E] = \xi = \delta_E^*(1_A) .$$

We shall show at the end of this section that the map Δ agrees with the classical map defined by means of factor sets.

We proceed with a number of assertions about naturality. Let $\alpha : A \to A'$ be a homomorphism of Q-modules, and let

$$E' : A' \rightarrowtail G' \twoheadrightarrow Q$$

be an extension with $\Delta[E'] = \xi' \in H^2(Q,A')$. Then we have

PROPOSITION 4.1. There exists $f' : G \to G'$ such that the diagram

$$\begin{array}{ccccccc} E & : & A & \rightarrowtail & G & \twoheadrightarrow & Q \\ & & \alpha\downarrow & & f'\downarrow & & \| \\ E' & : & A' & \rightarrowtail & G' & \twoheadrightarrow & Q \end{array} \tag{4.5}$$

is commutative if and only if $\alpha_*(\xi) = \xi' \in H^2(Q,A')$.

PROOF. Suppose first that (4.5) is commutative. Then (3.7) yields the commutative diagram

$$\begin{array}{ccc} \mathrm{Hom}_Q(A,A) & \xrightarrow{\delta^*_E} & H^2(Q,A) \\ \alpha_*\downarrow & & \alpha_*\downarrow \\ \mathrm{Hom}_Q(A,A') & \xrightarrow{\delta^*_E} & H^2(Q,A') \\ \alpha^*\uparrow & & \| \\ \mathrm{Hom}_Q(A',A') & \xrightarrow{\delta^*_{E'}} & H^2(Q,A') \end{array}$$

whence immediately $\alpha_*(\xi) = \xi'$.

Suppose now that $\alpha_*(\xi) = \xi'$. We have to show the existence of a map $f' : G \to G'$. To do so we construct an extension of Q by the Q-module A'

$$E_\alpha : A' \overset{h'}{\rightarrowtail} G_\alpha \overset{g'}{\twoheadrightarrow} Q .$$

We set $G_\alpha = A' \rtimes G/T$ where $T = \{(\alpha(-a),h(a))\}$. It is easy to see that T is a normal subgroup. The map h' is induced by the embedding $A' \to A' \rtimes G$ and g' is induced by $A' \rtimes G \to G \xrightarrow{g} Q$. It is then easy to check that

$$f_\alpha : G \to G_\alpha$$

induced by the embedding $G \to A' \wr G$ yields a commutative diagram

$$\begin{array}{rcccccc} E: & A & \rightarrowtail & G & \twoheadrightarrow & Q \\ & \alpha\downarrow & & f_\alpha\downarrow & & \| \\ E_\alpha: & A' & \rightarrowtail & G_\alpha & \twoheadrightarrow & Q . \end{array}$$

By the first part of Proposition 4.1. we have

$$\alpha_*(\xi) = \alpha_*(\Delta[E]) = \Delta[E_\alpha]$$

so that $\Delta[E_\alpha] = \xi' = \Delta[E']$. It follows that E_α is equivalent to E' , so that we indeed obtain a map

$$f' : G \xrightarrow{f_\alpha} G_\alpha \to G'$$

with the required properties.

Notice that we do not claim that f' is unique; in fact there will be many maps f' with the required properties, in general (see Propositions 4.3 and V.6.1.).

We finally remark that the square

$$\begin{array}{ccc} A & \rightarrowtail & G \\ \alpha\downarrow & & f_\alpha\downarrow \\ A' & \rightarrowtail & G_\alpha \end{array}$$

satisfies an obvious universal property. Details are left to the reader. Of course this universal property could be used in the proof of Proposition 4.1.

Let $\ell : \bar{Q} \to Q$ be a homomorphism and let

$$\bar{E} : A \rightarrowtail \bar{G} \twoheadrightarrow \bar{Q}$$

be an extension with $\Delta[\bar{E}] = \bar{\xi} \in H^2(\bar{Q},A)$. Then we have

PROPOSITION 4.2. *There exists* $\bar{f} : \bar{G} \to G$ *such that the diagram*

$$\begin{array}{lccccc} & \bar{E} : & A \rightarrowtail & \bar{G} & \twoheadrightarrow & \bar{Q} \\ (4.6) & & \| & \downarrow \bar{f} & & \downarrow \ell \\ & E : & A \rightarrowtail & G & \twoheadrightarrow & Q \end{array}$$

is commutative if and only if $\ell^*(\xi) = \bar{\xi}$.

PROOF: First suppose that (4.6) is commutative. Then (3.7) yields a commutative diagram

$$\begin{array}{ccc} \mathrm{Hom}_Q(A,A) & \xrightarrow{\delta_E^*} & H^2(Q,A) \\ \| & & \downarrow \ell^* \\ \mathrm{Hom}_{\bar{Q}}(A,A) & \xrightarrow{\delta_{\bar{E}}^*} & H^2(\bar{Q},A) \end{array}$$

whence immediately $\ell^*(\xi) = \bar{\xi}$.

Suppose now that $\ell^*(\xi) = \bar{\xi}$. We have to show the existence of a map $\bar{f} : \bar{G} \to G$. To do so we construct an extension

$$E^\ell : A \overset{\bar{h}}{\rightarrowtail} G^\ell \overset{\bar{g}}{\twoheadrightarrow} \bar{Q} .$$

Here G^ℓ is the subgroup of $G \times \bar{Q}$ consisting of all (x,y) , $x \in G$, $y \in \bar{Q}$ with $g(x) = \ell(y)$. The maps $\bar{h}$ is induced by $A \to G \to G \times \bar{Q}$ and $\bar{g}$ is induced by the projection $G \times \bar{Q} \to \bar{Q}$. It is easy to see that $f^\ell : G^\ell \to G$ induced by the projection $G \times \bar{Q} \to G$ yields a commutative diagram

$$\begin{array}{lcccc} E^\ell : & A \rightarrowtail & G^\ell & \twoheadrightarrow & \bar{Q} \\ & \| & \downarrow f^\ell & & \downarrow \ell \\ E : & A \rightarrowtail & G & \twoheadrightarrow & Q \end{array}$$

By the first part of Proposition 4.2 we have

$$\ell^*(\xi) = \ell^*(\Delta[E]) = \Delta[E^\ell]$$

so that $\Delta[E^\ell] = \bar{\xi} = \Delta[\bar{E}]$. It follows that $\bar{E}$ is equivalent to E^ℓ , so that we indeed obtain a map

$$\bar{f} : \bar{G} \to G^\ell \xrightarrow{f^\ell} G$$

with the required properties.

Again $\bar{f}$ will not be uniquely determined, in general (see Proposition 4.3). We remark that G^ℓ is nothing else but the pull-back of the maps $\ell : \bar{Q} \to Q$ and $G \twoheadrightarrow Q$. Of course the universal property of G^ℓ could be used in the proof of Proposition 4.2.

<u>PROPOSITION 4.3</u>. <u>Let</u>

$$\begin{array}{ccccccc} \bar{E} & : & A & \rightarrowtail & \bar{G} & \twoheadrightarrow & \bar{Q} \\ & & \alpha\downarrow & & & & \downarrow\ell \\ E' & : & A' & \rightarrowtail & G' & \twoheadrightarrow & Q \end{array}$$

<u>be given. There exists</u> $f : \bar{G} \to G'$ <u>making the diagram commutative if and only if</u>

(4.7) $$\alpha_*(\Delta[\bar{E}]) = \ell^*(\Delta[E']) .$$

<u>Moreover, if (4.7) holds, the homomorphisms</u> $f : \bar{G} \to G'$ <u>are in one-to-one correspondence with the derivations</u> $d : \bar{Q} \to A'$.

<u>PROOF</u>: If f exists then clearly (4.7) holds. To prove the converse we use the procedures of the proofs of Propositions 4.1 and 4.2 to construct the diagram

$$
\begin{array}{ccccccc}
\bar{E} & : & A & \rightarrowtail & \bar{G} & \twoheadrightarrow & \bar{Q} \\
 & & \alpha\downarrow & & \downarrow & & \| \\
\bar{E}_\alpha & : & A' & \rightarrowtail & \bar{G}_\alpha & \twoheadrightarrow & \bar{Q} \\
 & & \| & & & & \| \\
E'^{\ell} & : & A' & \rightarrowtail & G'^{\ell} & \twoheadrightarrow & \bar{Q} \\
 & & \| & & \downarrow & & \downarrow \ell \\
E' & : & A' & \rightarrowtail & G' & \twoheadrightarrow & Q
\end{array}
$$

Now we have, by (4.7),

$$\Delta[\bar{E}_\alpha] = \alpha_*(\Delta[\bar{E}]) = \ell^*(\Delta[E']) = \Delta[E'^{\ell}] ,$$

so that $\bar{E}_\alpha$ is equivalent to E'^{ℓ}. Then f may be defined as composition

$$f : \bar{G} \to \bar{G}_\alpha \to G'^{\ell} \to G' .$$

If $d : \bar{Q} \to A'$ is a derivation, then $f_1 : \bar{G} \to G'$ defined by $f_1(x) = d'(x)\cdot f(x)$, $x \in \bar{G}$ where d' is defined by

$$d' : \bar{G} \to \bar{Q} \xrightarrow{d} A' \to G'$$

is again a group homomorphism inducing α and ℓ. To prove this, let $x,y \in \bar{G}$. Then

$$
\begin{aligned}
(4.8)\quad f_1(x\cdot y) &= d'(x\cdot y)\cdot f(x\cdot y) = d'(x)\cdot[f(x)\cdot d'(y)\cdot(f(x))^{-1}]\cdot f(x)\cdot f(y) \\
&= f_1(x)\cdot f_1(y) .
\end{aligned}
$$

It is clear that f_1 induces α and ℓ. Conversely, if $f,f_1:\bar{G} \to G'$ are two homomorphisms inducing α and ℓ then the calculation (4.8) shows that $d'(x) = f_1(x)\cdot(f(x))^{-1}$ induces a derivation $d : \bar{Q} \to A'$. The proof of Proposition 4.3 is thus complete.

Finally we recall the description of extensions (with abelian kernel) by means of factor sets and relate this description with our map Δ as defined by (4.4). We consider the extension

(4.9) $$E : A \overset{h}{\rightarrowtail} G \overset{g}{\twoheadrightarrow} Q$$

with abelian kernel. Let $s : Q \to G$ be a section, i.e. a function with $gs = 1_Q$ and $s(e) = e$. Note that (sx), $x \in Q$ is just a set of representatives of Q in G. Define a function $\varphi : Q\times Q \to A$ by

(4.10) $$\varphi(x,y) = sxsy(s(xy))^{-1} \quad , \quad x,y \in Q .$$

The function φ is called a factor set. It may be interpreted as an element in $\mathrm{Hom}_Q(B_2',A)$ where $\underline{B}'$ denotes the normalized standard resolution in inhomogeneous form (see [43], p.216-217). It is easy to show that φ is a cocycle and that different sections s yield cohomologous cocycles, so that the cohomology class $[\varphi] \in H^2(Q,A)$ is well defined by (4.10) (see [57], p.111).

PROPOSITION 4.4. $[\varphi] = \Delta[E]$.

PROOF: Consider the exact sequence

$$A \overset{\kappa}{\rightarrowtail} \mathbb{Z}Q \otimes_G IG \overset{\nu}{\twoheadrightarrow} IQ$$

associated with the extension (4.9) (see [43], Theorem VI.6.3). We may then construct homomorphisms φ_1, φ_2 such that the diagram with exact rows

(4.11) $$\begin{array}{ccccc} B_2' & \overset{\partial_2}{\longrightarrow} & B_1' & \overset{\partial_1}{\twoheadrightarrow} & IQ \\ \downarrow\varphi_2 & & \downarrow\varphi_1 & & \| \\ A & \overset{\kappa}{\rightarrowtail} & \mathbb{Z}Q\otimes_G IG & \overset{\nu}{\twoheadrightarrow} & IQ \end{array}$$

is commutative. Note that the existence of φ_1 , φ_2 is asserted by the comparison theorem. They may be chosen to be

$$\varphi_1[x] = 1 \otimes (sx-1) \quad , \quad x \in Q \quad , \tag{4.12}$$

$$\varphi_2[x|y] = sxsy(s(xy))^{-1} = \varphi(x,y) \quad , \quad x,y \in Q \ . \tag{4.13}$$

To prove commutativity, consider

$$\begin{aligned}(\varphi_1\partial_2-\kappa\varphi_2)[x|y] &= 1 \otimes [sx(sy-1)-(s(xy)-1)+(sx-1)-(sxsy(s(xy))^{-1}-1)] \\ &= 1 \otimes [(sxsy(s(xy))^{-1}-1)s(xy)-(sxsy(s(xy))^{-1}-1)] \\ &= 1 \otimes [(sxsy(s(xy))^{-1}-1)(s(xy)-1)] \ .\end{aligned}$$

But this is zero, since $(sxsy(s(xy))^{-1}-1)$ may be moved to the left hand side, where it operates as zero. Applying $\mathrm{Hom}_Q(-,A)$ to diagram (4.11) we obtain

$$\begin{array}{ccccccc} \ldots \to \mathrm{Hom}_Q(ZQ\otimes_G IG,A) & \xrightarrow{h^*} & \mathrm{Hom}_Q(A,A) & \xrightarrow{\delta_E^*} & H^2(Q,A) & \to \ldots \\ \downarrow \varphi_1^* & & \downarrow \varphi_2^* & & \parallel & \\ \ldots \to \mathrm{Hom}_Q(B_1',A) & \xrightarrow{\partial_2^*} & \mathrm{Hom}_Q(B_2',A) & \xrightarrow{[\ ]} & H^2(Q,A) & \to 0 \end{array} \tag{4.14}$$

where the upper sequence is part of the 5-term sequence (3.2). From (4.14) it is then clear that

$$\Delta[E] = \delta_E^*(1_A) = [\varphi_2^*(1_A)] = [\varphi] \in H^2(Q,A) \quad ,$$

thus proving Proposition 4.4.

II.5. Universal Coefficients; the Künneth Theorem

Let C be an abelian group, regarded as trivial left or right Q-module. Then there are natural exact sequences ([43], p.222)

$$0 \to \mathrm{Ext}(H_{n-1}Q,C) \xrightarrow{\Sigma} H^n(Q,C) \xrightarrow{\Pi} \mathrm{Hom}(H_nQ,C) \to 0 , \tag{5.1}$$

$$(5.2)\qquad 0 \to H_nQ \otimes C \xrightarrow{\Pi'} H_n(Q,C) \xrightarrow{\Sigma'} \mathrm{Tor}(H_{n-1}Q,C) \to 0 .$$

These sequences are called the <u>universal coefficient exact sequences</u>. Both of them split, but the splitting is non-natural, in general. For our applications we shall need a generalization of the sequences (5.1), (5.2) in dimension $n = 2$.

<u>PROPOSITION 5.1</u>. <u>Let</u> q <u>be a non-negative integer and let</u> C <u>be a</u> Z/qZ-<u>module. Then the following sequences are exact and natural</u>.

$$(5.3)\quad 0 \to \mathrm{Ext}^1_{Z/qZ}(Q^q_{ab},C) \xrightarrow{\Sigma} H^2(Q,C) \xrightarrow{\Pi} \mathrm{Hom}(H^q_2Q,C) \to \mathrm{Ext}^2_{Z/qZ}(Q^q_{ab},C) \to \dots$$

$$(5.4)\quad \dots \to \mathrm{Tor}_2^{Z/qZ}(Q^q_{ab},C) \to Q^q_{ab} \otimes C \xrightarrow{\Pi'} H_2(Q,C) \xrightarrow{\Sigma'} \mathrm{Tor}_1^{Z/qZ}(Q^q_{ab},C) \to 0$$

<u>PROOF</u>: We only prove (5.3), the proof of (5.4) being dual. Let $S \overset{h'}{\rightarrowtail} F \overset{g'}{\twoheadrightarrow} Q$ be a free presentation of Q and let

$$0 \to H^q_2Q \xrightarrow{\delta_*} S/F\#_qS \xrightarrow{h^!_*} F^q_{ab} \xrightarrow{g^!_*} Q^q_{ab} \to 0$$

be the associated homology 5-term sequence with coefficients Z/qZ (see (3.13)). Set $K = \mathrm{im}\, h^!_*$, so that we have short exact sequences of Z/qZ-modules

$$(5.5)\qquad 0 \to H^q_2Q \to S/F\#_qS \to K \to 0 ,$$

$$(5.6)\qquad 0 \to K \to F^q_{ab} \to Q^q_{ab} \to 0 .$$

We apply $\mathrm{Hom}_{Z/qZ}(-,C)$ to these sequences. Using the abbreviations $\mathrm{Hom} = \mathrm{Hom}_{Z/qZ}$, $\mathrm{Ext}^i = \mathrm{Ext}^i_{Z/qZ}$ we obtain

$$(5.7)\quad 0 \to \mathrm{Hom}(Q^q_{ab},C) \to \mathrm{Hom}(F^q_{ab},C) \to \mathrm{Hom}(K,C) \to \mathrm{Ext}^1(Q^q_{ab},C) \to 0 ,$$

$$(5.8)\quad 0 \to \mathrm{Hom}(K,C) \to \mathrm{Hom}(S/F\#_qS,C) \to \mathrm{Hom}(H^q_2Q,C) \to \mathrm{Ext}^1(K,C) \to \dots$$

where we have a zero at the right hand end of (5.7) since F^q_{ab} is free over Z/qZ . Using (5.1) for $n = 1$ we have

$$\begin{aligned} \mathrm{Hom}(Q^q_{ab},C) &= \mathrm{Hom}_Z(Q_{ab},C) = H^1(Q,C) \quad , \\ \mathrm{Hom}(F^q_{ab},C) &= \mathrm{Hom}_Z(F_{ab},C) = H^1(F,C) . \end{aligned}$$

Also, it is easy to see that $\mathrm{Hom}(S/F\#_q S,C) = \mathrm{Hom}_Q(S_{ab},C)$. Furthermore it follows from the long exact $\mathrm{Ext}_{Z/qZ}$-sequence that

$$\mathrm{Ext}^1(K,C) \cong \mathrm{Ext}^2(Q^q_{ab},C) \quad ,$$

since F^q_{ab} is Z/qZ-free. We may compile the above information in the following diagram

$$\begin{array}{ccccccccccc}
 & & & & & & 0 & & & & \\
 & & & & & & \downarrow & & & & \\
0 & \to & \mathrm{Hom}(Q^q_{ab},C) & \to & \mathrm{Hom}(F^q_{ab},C) & \to & \mathrm{Hom}(K,C) & \to & \mathrm{Ext}^1(Q_{ab},C) & \to & 0 \\
 & & \| & & \| & & \downarrow & & \downarrow \Sigma & & \\
0 & \to & H^1(Q,C) & \to & H^1(F,C) & \to & \mathrm{Hom}_Q(S_{ab},C) & \xrightarrow{\delta^*} & H^2(Q,C) & \to & 0 \\
 & & & & & & \downarrow & & \downarrow \Pi & & \\
 & & & & & & \mathrm{Hom}(H^q_2Q,C) & = & \mathrm{Hom}(H^q_2Q,C) & & \\
 & & & & & & \downarrow & & \downarrow & & \\
 & & & & & & \mathrm{Ext}^1(K,C) & \cong & \mathrm{Ext}^2(Q^q_{ab},C) & & \\
 & & & & & & \downarrow & & \downarrow & & \\
 & & & & & & \vdots & & \vdots & &
\end{array} \tag{5.9}$$

where the second line is the 5-term cohomology sequence (see (3.3)). It is trivial that Σ is monomorphic and that the right most column is exact. The proof of (5.3) is thus complete.

We are of course interested in the case where

$$\mathrm{Ext}^2_{Z/qZ}(Q^q_{ab},C) \cong \mathrm{Ext}^1_{Z/qZ}(K,C) = 0$$

$$\mathrm{Tor}_2^{\mathbb{Z}/q\mathbb{Z}}(Q_{ab}^q, C) \cong \mathrm{Tor}_1^{\mathbb{Z}/q\mathbb{Z}}(K, C) = 0$$

for then sequences (5.3), (5.4) become short exact. This is so, if, for example K is $\mathbb{Z}/q\mathbb{Z}$-projective. However we have

LEMMA 5.2. *Let* $q \neq 0$. *The* $\mathbb{Z}/q\mathbb{Z}$-*module* K *is projective if and only if* Q_{ab}^q *is.*

PROOF: It is well-known that $\mathbb{Z}/q\mathbb{Z}$ is self-injective, and (of course) noetherian. It follows that every free $\mathbb{Z}/q\mathbb{Z}$-module is injective. Now if K is projective it is a direct summand in a free, and hence injective module. Thus K is injective and sequence (5.6) splits. As a direct summand in a free module Q_{ab}^q is projective. Conversely, if Q_{ab}^q is projective, sequence (5.6) splits, and K is projective, also.

COROLLARY 5.3. *Let* $q \geq 1$ *and let* C *be a* $\mathbb{Z}/q\mathbb{Z}$-*module. If* Q_{ab}^q *is* $\mathbb{Z}/q\mathbb{Z}$-*projective, then*

(5.10)
$$H^2(Q,C) \cong \mathrm{Hom}(H_2^q Q, C) ,$$
$$H_2(Q,C) \cong H_2^q Q \otimes C .$$

The following proposition yields some additional information on the homomorphism Π in (5.3). Let $E : N \rightarrowtail G \twoheadrightarrow Q$ be an extension with N central and N a $\mathbb{Z}/q\mathbb{Z}$-module. Then the homology 5-term sequence (3.13) of E reads

(5.11)
$$H_2^q G \to H_2^q Q \xrightarrow{\delta_*^E} N \to G_{ab}^q \to Q_{ab}^q \to 0 .$$

We then have

PROPOSITION 5.4. $\Pi(\Delta[E]) = \delta_*^E : H_2^q Q \to N$.

PROOF: We consider a free presentation $E' : S \overset{h'}{\rightarrowtail} F \overset{g'}{\twoheadrightarrow} Q$ of Q and construct the diagram

$$\begin{array}{lccccc} E' : & S & \rightarrowtail & F & \twoheadrightarrow & Q \\ (5.12) & s\downarrow & & \downarrow & & \| \\ E : & N & \rightarrowtail & G & \twoheadrightarrow & Q \end{array}$$

By (3.7) it gives rise to a commutative diagram of 5-term sequences

$$\begin{array}{ccccccccc} 0 & \to & H_2^q Q & \to & S/F\#_q S & \xrightarrow{h_*^!} & F_{ab}^q & \to & Q_{ab}^q \to 0 \\ \downarrow & & \| & & s'\downarrow & & \downarrow & & \| \\ H_2^q G & \to & H_2^q Q & \xrightarrow{\delta_*^E} & N & \to & G_{ab}^q & \to & Q_{ab}^q \to 0 \end{array}$$

It follows that

$$\delta_*^E = s'|\ker(h_*^! : S/F\#_q S \to F_{ab}^q) .$$

We may then read off from diagram (5.9) that

(5.11) $$\Pi\delta_{E'}^*(s') = \delta_*^E .$$

It remains to show that $\delta_{E'}^*(s') = \Delta[E]$. To do so we consider the diagram of 5-term sequences in cohomology with coefficients N , arising from (5.10)

$$\begin{array}{ccc} \mathrm{Hom}_Q(N,N) & \xrightarrow{\delta_E^*} & H^2(Q,N) \\ s^*\downarrow & & \| \\ \mathrm{Hom}_Q(S_{ab},N) & \xrightarrow{\delta_{E'}^*} & H^2(Q,N) . \end{array}$$

But now $\mathrm{Hom}_Q(S_{ab},N) \cong \mathrm{Hom}(S/F\#_q S,N)$, so that we indeed obtain

(5.12) $$\Delta[E] = \delta_E^*(1_N) = \delta_{E'}^* s^*(1_N) = \delta_{E'}^*(s') .$$

Thus the proof is complete.

Next we state the <u>Künneth Theorem</u>. Let G_1, G_2 be two groups and let $G = G_1 \times G_2$ be their direct product. Then there is a natural exact sequence (Künneth-sequence; see [43], p.223)

$$(5.13)\qquad 0 \to \bigoplus_{i+k=n} H_iG_1 \otimes H_kG_2 \to H_nG \to \bigoplus_{i+k=n-1} \mathrm{Tor}(H_iG_1, H_kG_2) \to 0 .$$

The Künneth sequence splits, but non-naturally, in general.

II.6. The Mayer-Vietoris Sequence and the Coproduct Theorem

In this section we first consider the (co)homology of a free product with amalgamated subgroup. We will then specialize to free products in order to obtain the coproduct theorem. Also, we will generalize the coproduct theorem to arbitrary index sets.

Let G_1, G_2 be two groups and let U be a subgroup of G_1 and G_2. Denote by $\bar{G}$ the free product of G_1 and G_2 with amalgamated subgroup U. It is well-known that

$$(6.1)\qquad \begin{array}{ccc} U & \overset{h_1}{\rightarrowtail} & G_1 \\ {\scriptstyle h_2}\downarrow & & \downarrow{\scriptstyle g_1} \\ G_2 & \overset{g_2}{\longrightarrow} & \bar{G} \end{array}$$

is a push-out diagram in $\underline{\underline{Gr}}$. Also, it is well-known that the maps $g_i : G_i \to \bar{G}$ and hence $g_ih_i : U \to \bar{G}$ are injective.

PROPOSITION 6.1. *Let A be a left $\bar{G}$-module and let B be a right $\bar{G}$-module. Then there are exact sequences* (Mayer-Vietoris)

$$(6.2)\qquad \begin{array}{l} 0 \to H^0(\bar{G},A) \xrightarrow{g^*} H^0(G_1,A) \oplus H^0(G_2,A) \xrightarrow{h^*} H^0(U,A) \to H^1(\bar{G},A) \to \cdots \\ \quad \cdots \to H^n(\bar{G},A) \xrightarrow{g^*} H^n(G_1,A) \oplus H^n(G_2,A) \xrightarrow{h^*} H^n(U,A) \to H^{n+1}(\bar{G},A) \to \cdots \end{array}$$

$$(6.3)\qquad \begin{array}{l} \cdots \to H_n(U,B) \xrightarrow{h_*} H_n(G_1,B) \oplus H_n(G_2,B) \xrightarrow{g_*} H_n(\bar{G},B) \to H_{n-1}(U,B) \to \cdots \\ \quad \cdots \to H_1(\bar{G},B) \to H_0(U,B) \xrightarrow{h_*} H_0(G_1,B) \oplus H_0(G_2,B) \xrightarrow{g_*} H_0(\bar{G},B) \to 0 \end{array}$$

<u>where</u> $g^* = \{g_1^*, g_2^*\}$, $h^* = \langle h_1^*, -h_2^*\rangle$; <u>and</u> $g_* = \langle g_{1*}, g_{2*}\rangle$, $h_* = \{h_{1*}, -h_{2*}\}$.

<u>PROOF</u>: We first show that

$$(6.4)\qquad \begin{array}{ccc} \mathbb{Z}\bar{G} \otimes_U IU & \xrightarrow{h_{1*}} & \mathbb{Z}\bar{G} \otimes_{G_1} IG_1 \\ h_{2*}\downarrow & & g_{1*}\downarrow \\ \mathbb{Z}\bar{G} \otimes_{G_2} IG_2 & \xrightarrow{g_{2*}} & I\bar{G} \end{array}$$

is both a pull-back and a push-out square in $\underline{\underline{\mathrm{Mod}}}_{\bar{G}}$. In order to prove the pull-back property it is enough to show that h_{i*} in (6.4) is monomorphic. But we have

$$\mathbb{Z}\bar{G} \otimes_U IU = \mathbb{Z}\bar{G} \otimes_{G_i} (\mathbb{Z}G_i \otimes_U IU) .$$

Since U is a subgroup of G_i the map $\mathbb{Z}G_i \otimes_U IU \to IG_i$ is monomorphic; since G_i is a subgroup of $\bar{G}$, tensoring with $\mathbb{Z}\bar{G}$ over G_i shows that h_{i*} is monomorphic, also.

In order to prove that (6.4) is a push-out square, we verify the universal property. Thus let M be any $\bar{G}$-module and let $\alpha_i : \mathbb{Z}\bar{G} \otimes_{G_i} IG_i \to M$ with $\alpha_1 h_{1*} = \alpha_2 h_{2*}$ be given. By (II.2.2) the maps α_1, α_2 yield a (unique) pair of derivations $d_i : G_i \to M$ with

$$(6.5)\qquad d_1 h_1 = d_2 h_2 : U \to M .$$

By (II.2.3) the derivations d_1, d_2 correspond to group homomorphisms $f_i : G_i \to M \rtimes G_i$ with $G_i \to M \rtimes G_i \to G_i$ the identity. Of course, the square

$$\begin{array}{ccccc}
U & \xrightarrow{h_1} & G_1 & \xrightarrow{f_1} & M \rtimes G_1 \\
{\scriptstyle h_2}\downarrow & & & & \\
G_2 & & & & \downarrow \\
{\scriptstyle f_2}\downarrow & & & & \\
M \rtimes G_2 & & \longrightarrow & & M \rtimes \bar{G}
\end{array}$$

is commutative. Since (6.1) is a push-out square we obtain a (unique) map $f : \bar{G} \to M \rtimes \bar{G}$. Also, it follows from the uniqueness part of the push-out property of (6.1) that $\bar{G} \to M \rtimes \bar{G} \to \bar{G}$ is the identity. Applying (II.2.3) and (II 2.2) again we obtain a (unique) map $\alpha : I\bar{G} \to M$ satisfying

$$\alpha_i = \alpha g_{i*} : Z\bar{G} \otimes_{G_i} IG_i \to M \quad , \quad i = 1,2 \; .$$

Thus (6.4) is both a pull-back and a push-out square. Now consider the diagram

$$\begin{array}{ccccccc}
0 \to & Z\bar{G}\otimes_U IU & \xrightarrow{\{h_{1*},-h_{2*}\}} & Z\bar{G}\otimes_{G_1} IG_1 \oplus Z\bar{G}\otimes_{G_2} IG_2 & \xrightarrow{<g_{1*},g_{2*}>} & I\bar{G} & \to 0 \\
 & \downarrow & & \downarrow & & \downarrow & \\
(6.6) \quad 0 \to & Z\bar{G}\otimes_U ZU & \xrightarrow{\{h_{1*},-h_{2*}\}} & Z\bar{G}\otimes_{G_1} ZG_1 \oplus Z\bar{G}\otimes_{G_2} ZG_2 & \xrightarrow{<g_{1*},g_{2*}>} & Z\bar{G} & \to 0 \\
 & \downarrow & & \downarrow & & \downarrow & \\
0 \to & Z\bar{G}\otimes_U Z & \xrightarrow{\{h_{1*},-h_{2*}\}} & Z\bar{G}\otimes_{G_1} Z \oplus Z\bar{G}\otimes_{G_2} Z & \xrightarrow{<g_{1*},g_{2*}>} & Z & \to 0
\end{array}$$

Obviously all columns are exact. Since (6.4) is a pull-back and a push-out the top row is exact. The second row is exact, since it is isomorphic to

$$0 \to Z\bar{G} \xrightarrow{\{1,-1\}} Z\bar{G} \oplus Z\bar{G} \xrightarrow{<1,1>} Z\bar{G} \to 0 \; .$$

It follows that the bottom row is exact, also. Now apply the functors $\mathrm{Hom}_{\bar{G}}(-,A)$ and $B \otimes_{\bar{G}} -$ to the bottom row of (6.6). Using that for any subgroup V of $\bar{G}$ we have

(6.7) $$\mathrm{Ext}^n_{\bar{G}}(Z\bar{G} \otimes_V Z,A) = \mathrm{Ext}^n_V(Z,A) = H^n(V,A) \;;$$

(6.8) $$\mathrm{Tor}^{\bar{G}}_n(B,Z\bar{G} \otimes_V Z) = \mathrm{Tor}^V_n(B,Z) = H_n(V,B) \;;$$

we conclude that the resulting long exact sequences are just (6.2) and (6.3).

<u>COROLLARY 6.2</u>. <u>Let</u> $\bar{G} = G_1 * G_2$ <u>be the free product of</u> G_1 <u>and</u> G_2 ; <u>let</u> A <u>be a left</u> $\bar{G}$-<u>module, and let</u> B <u>be a right</u> $\bar{G}$-<u>module. Then the coproduct injections</u> $g_i : G_i \to \bar{G}$, $i = 1,2$ <u>induce isomorphisms</u>

(6.9) $$g^* : H^n(\bar{G},A) \cong H^n(G_1,A) \oplus H^n(G_2,A) \quad , \quad n \geq 2 \;;$$

(6.10) $$g_* \quad H_n(G_1,B) \oplus H_n(G_2,B) \cong H_n(\bar{G},B) \quad , \quad n \geq 2 \;.$$

<u>Moreover the sequences</u>

$$0 \to H^0(\bar{G},A) \xrightarrow{g^*} H^0(G_1,A) \oplus H^0(G_2,A) \to A \to$$

$$\to H^1(\bar{G},A) \xrightarrow{g^*} H^1(G_1,A) \oplus H^1(G_2,A) \to 0$$

$$0 \to H_1(G_1,B) \oplus H_1(G_2,B) \xrightarrow{g_*} H_1(\bar{G},B) \to$$

$$\to B \to H_0(G_1,B) \oplus H_0(G_2,B) \xrightarrow{g_*} H_0(\bar{G},B) \to 0$$

<u>are exact</u>.

<u>PROOF</u>: Use (6.2), (6.3) for $U = e$ and note that

$$H^n(e,A) = 0 = H_n(e,B) \qquad \text{for} \quad n \geq 1 \;.$$

We remark that if A and B are <u>trivial</u> $\bar{G}$-modules, the assertion of Corollary 6.2 is true for $n = 1$, also. We leave the obvious proof of this fact to the reader. We finally note that there are generalizations of both Proposition 6.1 and Corollary 6.2 to arbitrary index sets. We will be content to state and prove explicitly the generalization of Corollary 6.2.

PROPOSITION 6.3. Let $\bar{G} = \underset{i\in I}{\bigstar} G_i$ be the free product of (G_i), $i \in I$. Let A be a left $\bar{G}$-module, and let B be a right $\bar{G}$-module. Then the coproduct injections $g_i : G_i \to \bar{G}$, $i \in I$ induce isomorphisms

$$(6.11) \qquad g^* : H^n(\bar{G},A) \tilde{\to} \prod_{i\in I} H^n(G_i,A) \qquad n \geq 2 ;$$

$$(6.12) \qquad g_* : \bigoplus_{i\in I} H_n(G_i,B) \tilde{\to} H_n(\bar{G},B) \quad , \quad n \geq 2 .$$

PROOF: In order to keep our proofs as transparent as possible, we will repeat in the proof below some arguments already used in the proof of Proposition 6.1.

We first show that it is enough to prove that for any $\bar{G}$-module M the coproduct injections induce an isomorphism

$$(6.13) \qquad g^* : \mathrm{Der}(\bar{G},M) \tilde{\to} \prod_{i\in I} \mathrm{Der}(G_i,M) .$$

Given (6.13) we obtain, using (II.2.2),

$$\begin{aligned} \mathrm{Hom}_{\bar{G}}(I\bar{G},M) &\cong \prod_{i\in I} \mathrm{Hom}_{G_i}(IG_i,M) \\ &\cong \prod_{i\in I} \mathrm{Hom}_{\bar{G}}(\mathbb{Z}\bar{G} \otimes_{G_i} IG_i,M) \\ &= \mathrm{Hom}_{\bar{G}}(\bigoplus_{i\in I} \mathbb{Z}\bar{G} \otimes_{G_i} IG_i,M) . \end{aligned}$$

Hence we may conclude that the coproduct injections induce an isomorphism

$$(6.14) \qquad g_* : \bigoplus_{i\in I} \mathbb{Z}\bar{G} \otimes_{G_i} IG_i \tilde{\to} I\bar{G} .$$

Applying $\mathrm{Ext}_{\bar{G}}^{n-1}(-,A)$, $n \geq 2$ yields (6.11); applying $\mathrm{Tor}_{n-1}^{\bar{G}}(B,-)$, $n \geq 2$ yields (6.12).

We now prove (6.13). The homomorphism g^* is given by composing a derivation $d : \bar{G} \to M$ with the coproduct injections $g_i : G_i \to \bar{G}$. In order to construct an inverse of g^* let the derivations $d_i : G_i \to M$, $i \in I$ be given. By (II.2.3) each d_i corresponds to a group homomorphism $G_i \to M \rtimes G_i$ with $G_i \to M \rtimes G_i \to G_i$ the identity. Using the universal property of the free product we may conclude that the homomorphism $G_i \to M \rtimes G_i \to M \rtimes \bar{G}$ there is a unique homomorphism $\bar{G} \to M \rtimes \bar{G}$ with $\bar{G} \to M \rtimes \bar{G} \to \bar{G}$ the identity. Applying (II.2.3) we obtain a derivation $d : \bar{G} \to M$. It is easy to see that this procedure yields an inverse of g^* in (6.13). Thus the proof of Proposition 6.3 is complete.

CHAPTER III

EXTENSIONS IN $\underline{\underline{V}}$ WITH ABELIAN KERNEL

In this chapter we define the functors $\tilde{V}$, V associated with a variety $\underline{\underline{V}}$ and deduce their basic properties.

The definition of $\tilde{V}$, V is given in Section 1. In Section 2 we deduce exact sequences that are analogous to the 5-term sequences in ordinary (co)homology (see Section II.3). It becomes apparent from these sequences that the functors $\tilde{V}$, V correspond, at least formally, to the second (co)homology group functor. This point of view is substantiated in the important Section 3. It is shown that if the semi-direct product $A \rtimes Q$ is in $\underline{\underline{V}}$ then $\tilde{V}(Q,A)$ classifies extensions in $\underline{\underline{V}}$ in exactly the same way as $H^2(Q,A)$ classifies extensions in $\underline{\underline{Gr}}$. As a consequence it becomes clear that the natural domain of the functors $\tilde{V}(Q,-)$ (and $V(Q,-)$) is not $\underline{\underline{Mod}}_Q$ but the full subcategory $\underline{\underline{VMod}}_Q$ consisting of those Q-modules A for which the semi-direct product $A \rtimes Q$ is in $\underline{\underline{V}}$. In Section 4 various characterizations of $\underline{\underline{VMod}}_Q$ and examples are given.

The remaining sections of this chapter deal with the functors $\tilde{V}$, V in greater detail. In an elementary manner we establish their relevant properties. In Section 5 we prove the existence of coefficient sequences that correspond to the long exact (co)homology sequences in ordinary (co)homology. In Section 6 we establish for $\tilde{V}$, V the analog of the coproduct theorem (Corollary II.6.2). In Section 7 we deduce a change of variety sequence, and finally, in Section 8 we prove universal coefficient sequences for $\tilde{V}$, V .

Most of the results presented in this chapter are well known. In particular, we want to mention the following papers: André [1] (Sections 2, 5), Barr-Beck [9] (Sections 2, 5), Knopfmacher [50] (Sections 3, 4), Leedham-Green [53], [54], [55] (Sections 2, 4, 5, 7, 8). Ideas leading up to our elementary approach are to be found in Stammbach [78], [79].

III.1. The Groups $\tilde{V}(Q,A)$, $V(Q,B)$

Let $\underline{\underline{V}}$ be an arbitrary variety, and let Q be a group in $\underline{\underline{V}}$. By A we denote a left Q-module and by B we denote a right Q-module. Our intention is to define functors

(1.1) $$\tilde{V}(Q,-) : \underline{\underline{Mod}}^{\ell}{}_{Q} \to \underline{\underline{Ab}} ,$$

(1.2) $$V(Q,-) : \underline{\underline{Mod}}^{r}{}_{Q} \to \underline{\underline{Ab}} .$$

Their values on modules A,B are given as follows. Let $f : F \twoheadrightarrow Q$ be a $\underline{\underline{V}}$-free presentation of Q, then

(1.3) $$\tilde{V}(Q,A) = \ker(f^*: H^2(Q,A) \to H^2(F,A)) ,$$

(1.4) $$V(Q,B) = \operatorname{coker}(f_*: H_2(F,B) \to H_2(Q,B)) .$$

We have to show that these groups do not depend on the chosen presentation $f : F \twoheadrightarrow Q$. We do this for $\tilde{V}(Q,A)$ only, the proof for $V(Q,B)$ being dual.

Let $f' : F' \twoheadrightarrow Q$ be another $\underline{\underline{V}}$-free presentation of Q. Then there exist maps h,h' making the triangle

$$(1.5)\qquad \begin{array}{ccc} F & \underset{h}{\overset{h'}{\rightleftarrows}} & F' \\ & {\scriptstyle f}\searrow \quad \swarrow{\scriptstyle f'} & \\ & Q & \end{array}$$

commutative. It follows that

$$(1.6)\qquad \begin{array}{ccc} & H^2(Q,A) & \\ & {\scriptstyle f^*}\swarrow \quad \searrow{\scriptstyle f'^*} & \\ H^2(F,A) & \underset{h^*}{\overset{h'^*}{\rightleftarrows}} & H^2(F',A) \end{array}$$

is commutative. Hence $\ker f^* = \ker h^*f'^* \supseteq \ker f'^*$, and conversely $\ker f'^* \supseteq \ker f^*$, thus proving equality.

The effect of $\tilde{V}(Q,-)$, $V(Q,-)$ on homomorphisms is as follows. Let $\alpha : A \to A'$, $\beta : B \to B'$ be homomorphisms. Then the following diagrams define α_* and β_*.

$$(1.7)\qquad \begin{array}{ccccccc} 0 & \to & \tilde{V}(Q,A) & \to & H^2(Q,A) & \to & H^2(F,A) \\ & & \alpha_* \downarrow & & \alpha_* \downarrow & & \alpha_* \downarrow \\ 0 & \to & \tilde{V}(Q,A') & \to & H^2(Q,A') & \to & H^2(F,A') \end{array}$$

$$(1.8)\qquad \begin{array}{ccccccc} H_2(F,B) & \to & H_2(Q,B) & \to & V(Q,B) & \to & 0 \\ \beta_* \downarrow & & \beta_* \downarrow & & \beta_* \downarrow & & \\ H_2(F,B') & \to & H_2(Q,B') & \to & V(Q,B') & \to & 0 \end{array}$$

We remark that for $\underline{\underline{V}} = \underline{\underline{Gr}}$ we have

$$(1.9)\qquad \tilde{V}(Q,A) = H^2(Q,A) ,$$

$$(1.10)\qquad V(Q,B) = H_2(Q,B)$$

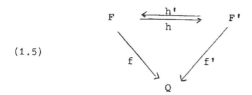

so that $\tilde{V}$ generalizes the second cohomology group functor and V the second homology group functor. We also note the obvious result

<u>PROPOSITION 1.1</u>. <u>For all $\underline{\underline{V}}$-free groups</u> F <u>and for all</u> F-<u>modules</u> A <u>and</u> B <u>we have</u>

$$\tilde{V}(F,A) = 0 = V(F,B) \ . \tag{1.11}$$

Next we study the behavior of $\tilde{V}$, V with respect to maps in the first variable. We restrict our discussion to $\tilde{V}$ leaving to the reader the dualization to V. Thus let $g : Q \to Q'$ be a homomorphism in $\underline{\underline{V}}$ and let A be a left Q'-module. Choose $\underline{\underline{V}}$-free presentations

$$f : F \twoheadrightarrow Q \ , \ f' : F' \twoheadrightarrow Q' \ .$$

The universal property of $\underline{\underline{V}}$-free groups yields a map $h : F \to F'$ such that the square

$$\begin{array}{ccc} F & \xrightarrow{f} & Q \\ h\downarrow & & \downarrow g \\ F' & \xrightarrow{f'} & Q \end{array}$$

is commutative. Hence we obtain the commutative diagram

$$\begin{array}{ccccccc} 0 \to & \tilde{V}(Q',A) & \to & H^2(Q',A) & \to & H^2(F',A) \\ & g^*\downarrow & & g^*\downarrow & & h^*\downarrow \\ 0 \to & \tilde{V}(Q,A) & \to & H^2(Q,A) & \to & H^2(F,A) \end{array} \tag{1.12}$$

with exact rows defining $g^* : \tilde{V}(Q',A) \to \tilde{V}(Q,A)$.

It is now obvious that $\tilde{V}(-,-)$ may be regarded as a functor on the full subcategory $\underline{\underline{VG}}^*$ of $\underline{\underline{G}}^*$ consisting of pairs (Q,A) with Q in $\underline{\underline{V}}$ and A a Q-module (see Section II.1). Again we leave it to the reader to formulate the analogous statement for $V(Q,B)$.

We finally remark the obvious, but important fact that for A a trivial module the definition of g^* in (1.12) makes $\tilde{V}(-,A)$ into a functor

$$(1.13) \qquad \tilde{V}(-,A) : \underline{\underline{V}} \to \underline{\underline{Ab}} .$$

Similarly, if B is a trivial module, we have a functor

$$(1.14) \qquad V(-,B) : \underline{\underline{V}} \to \underline{\underline{Ab}} .$$

III.2. The 5-Term Sequences

In this section we establish 5-term sequences for $\tilde{V}$ and V analogous to (II.3.2), ...,(II.3.5).

THEOREM 2.1. *Let* $E : N \overset{h}{\rightarrowtail} G \overset{g}{\twoheadrightarrow} Q$ *be an extension of groups with* G *in* $\underline{\underline{V}}$. *Let* A *be a left* Q-*module and let* B *be a right* Q-*module. Then there are exact sequences*

$$(2.1) \qquad 0 \to \mathrm{Der}(Q,A) \xrightarrow{g^*} \mathrm{Der}(G,A) \xrightarrow{h^*} \mathrm{Hom}_Q(N_{ab},A) \xrightarrow{\delta^*_E} \tilde{V}(Q,A) \xrightarrow{g^*} \tilde{V}(G,A) ,$$

$$(2.2) \qquad 0 \to H^1(Q,A) \xrightarrow{g^*} H^1(G,A) \xrightarrow{h^*} \mathrm{Hom}_Q(N_{ab},A) \xrightarrow{\delta^*_E} \tilde{V}(Q,A) \xrightarrow{g^*} \tilde{V}(G,A) ,$$

$$(2.3) \qquad V(G,B) \xrightarrow{g_*} V(Q,B) \xrightarrow{\delta^E_*} B\otimes_Q N_{ab} \xrightarrow{h_*} B\otimes_G IG \xrightarrow{g_*} B\otimes_Q IQ \to 0 ,$$

$$(2.4) \qquad V(G,B) \xrightarrow{g_*} V(Q,B) \xrightarrow{\delta^E_*} B\otimes_Q N_{ab} \xrightarrow{h_*} H_1(G,B) \xrightarrow{g_*} H_1(Q,B) \to 0 .$$

PROOF: Let $f : F \twoheadrightarrow G$ be a $\underline{\underline{V}}$-free presentation of G . Then $gf : F \twoheadrightarrow Q$ is a $\underline{\underline{V}}$-free presentation of Q . Using the 5-term sequence (II.3.2) we obtain the following commutative diagram with exact rows and columns.

$$
\begin{array}{cccccccccc}
 & & & & & & & 0 & & 0 \\
 & & & & & & & \downarrow & & \downarrow \\
 & & & & & & & \tilde{V}(Q,A) & \xrightarrow{g^*} & V(G,A) \\
 & & & & & & \nearrow & \downarrow & & \downarrow \\
(2.5)\quad 0 & \to & Der(Q,A) & \to & Der(G,A) & \to & Hom_Q(N_{ab},A) \xrightarrow{\delta^*_E} & H^2(Q,A) & \xrightarrow{g^*} & H^2(G,A) \\
 & & & & & & & (gf)^*\downarrow & & \downarrow f^* \\
 & & & & & & & H^2(F,A) & = & H^2(F,A)
\end{array}
$$

It is then obvious that $\delta^*_E : Hom_Q(N_{ab},A) \to H^2(Q,A)$ factors through $\tilde{V}(Q,A)$ and that sequence (2.1) is exact.

We leave it to the reader to formulate the exact statement of naturality (see (II.3.7)) and to give its proof.

Later on we shall need sequence (2.3) (or (2.4)) in the special case when $B = Z/qZ$. Using the notation introduced in Section II.3 and writing

$$(2.6)\qquad V^qG = V(G,Z/qZ)$$

we have the exact sequence

$$(2.7)\qquad V^qG \xrightarrow{g_*} V^qQ \xrightarrow{\delta^E_*} N/G \#_q N \xrightarrow{h_*} G^q_{ab} \xrightarrow{g_*} Q^q_{ab} \to 0 .$$

Next we note

<u>COROLLARY 2.2</u>. <u>Let</u> $R \rightarrowtail^{h} F \overset{g}{\twoheadrightarrow} Q$ <u>be a $\underline{\underline{V}}$-free presentation. Then</u>

$$(2.8)\qquad \tilde{V}(Q,A) = coker(h^*:Der(F,A) \to Hom_Q(R_{ab},A)) ,$$

$$(2.9)\qquad V(Q,B) = ker(h_*:B \otimes_Q R_{ab} \to B \otimes_F IF) .$$

<u>In particular</u>,

$$(2.10)\qquad V^qQ = (R \cap F \#_q F)/F \#_q R .$$

PROOF: This immediately follows from sequences (2.1), (2.3) using the fact that that $\tilde{V}(F,A) = 0 = V(F,B)$ (see 1.11).

We conclude with an example. Let $\underline{\underline{V}} = \underline{\underline{Ab}}$ and let Q be in $\underline{\underline{V}}$. Let C be a trivial Q-module. Choosing a $\underline{\underline{V}}$-free i.e. free abelian presentation $R \rightarrowtail F \twoheadrightarrow Q$ of Q we have exact sequences

$$0 \to \mathrm{Hom}(Q,C) \to \mathrm{Hom}(F,C) \to \mathrm{Hom}(R,C) \to \tilde{V}(Q,C) \to 0 \ , \tag{2.11}$$

$$0 \to V(Q,C) \to C \otimes R \to C \otimes F \to C \otimes Q \to 0 \ . \tag{2.12}$$

It is apparent from this that under the given hypotheses (C a trivial Q-module!) we have

$$\tilde{V}(Q,C) = \mathrm{Ext}^1_Z(Q,C) \ , \tag{2.13}$$

$$V(Q,C) = \mathrm{Tor}^Z_1(C,Q) \ . \tag{2.14}$$

III.3. The Group of Extensions

Let Q be a group in $\underline{\underline{V}}$ and let A be a Q-module. We want to describe those extensions of Q by A

$$E : A \overset{h}{\rightarrowtail} G \overset{g}{\twoheadrightarrow} Q \tag{3.1}$$

with G in $\underline{\underline{V}}$. If this is the case we shall say that the extension E is in $\underline{\underline{V}}$. It is clear that if E is in $\underline{\underline{V}}$, then every extension equivalent to E is in $\underline{\underline{V}}$, also. Note that if E is in $\underline{\underline{V}}$, then A as abelian group is in $\underline{\underline{V}}$, also.

Let $\Delta[E] = \xi \in H^2(Q,A)$ be the 2-cohomology class associated with E by (II.4.4). Let $f : F \twoheadrightarrow Q$ be a $\underline{\underline{V}}$-free presentation of Q. Define $\eta \in H^2(F,A)$ by

$$\eta = f^*(\xi)$$

where $f^* : H^2(Q,A) \to H^2(F,A)$. We may then state

<u>PROPOSITION 3.1</u>. <u>The extension</u> E <u>is in</u> $\underline{\underline{V}}$ <u>if and only if</u> $\eta = 0$ <u>and</u> $A \rtimes F$ <u>is in</u> $\underline{\underline{V}}$.

<u>PROOF</u>: Consider the diagram

$$\begin{array}{lccccc} & E^f : & A \rightarrowtail & G^f & \twoheadrightarrow & F \\ (3.2) & & \| & \downarrow & & \downarrow f \\ & E : & A \overset{h}{\rightarrowtail} & G & \overset{g}{\twoheadrightarrow} & Q \end{array}$$

Recall that G^f is defined by

$$G^f = \{(x,y) \in G \times F \mid gx = fy\} .$$

Now suppose that G is in $\underline{\underline{V}}$, then $G \times F$ is in $\underline{\underline{V}}$ and hence G^f is in $\underline{\underline{V}}$. The universal property of the $\underline{\underline{V}}$-free group F then yields a splitting $t : F \to G^f$ of the extension E^f . It follows that $G^f = A \rtimes F$, so that $\eta = f^*(\xi) = 0$ and $A \rtimes F$ is in $\underline{\underline{V}}$. Conversely suppose that $\eta = f^*(\xi) = 0$ and $A \rtimes F$ is in $\underline{\underline{V}}$. Then $G^f = A \rtimes F$, so that G^f is in $\underline{\underline{V}}$. But then, the group G , being an epimorphic image of G^f , is in $\underline{\underline{V}}$ also.

<u>COROLLARY 3.2</u>. <u>Let</u> $E : A \rightarrowtail G \twoheadrightarrow Q$ <u>be in</u> $\underline{\underline{V}}$; <u>then</u> $A \rtimes Q$ <u>is in</u> $\underline{\underline{V}}$.

<u>PROOF</u>: By Proposition 3.1 we conclude that for a $\underline{\underline{V}}$-free presentation $f : F \twoheadrightarrow Q$ the group $A \rtimes F$ is in $\underline{\underline{V}}$. Thus $A \rtimes Q$, being an epimorphic image of $A \rtimes F$, is in $\underline{\underline{V}}$ also.

THEOREM 3.3. *Let* Q *be in* $\underline{\underline{V}}$ *and let* A *be a* Q-*module*.

(i) *If* $A \rtimes Q$ *is not in* $\underline{\underline{V}}$, *then every extension* $E : A \rightarrowtail G \twoheadrightarrow Q$ *is outside* $\underline{\underline{V}}$.

(ii) *If* $A \rtimes Q$ *is in* $\underline{\underline{V}}$, *then the extensions* $E : A \rightarrowtail G \twoheadrightarrow Q$ *in* $\underline{\underline{V}}$ *are classified by* $\tilde{V}(Q,A)$.

PROOF: Assertion (i) is clear from Corollary 3.2. For the statement (ii) we proceed as follows. Let $f : F \twoheadrightarrow Q$ be a $\underline{\underline{V}}$-free presentation. Since $A \rightarrowtail A \rtimes Q \twoheadrightarrow Q$ is in $\underline{\underline{V}}$ it follows by Proposition 3.1 that $A \rtimes F$ is in $\underline{\underline{V}}$. We then conclude that E is in $\underline{\underline{V}}$ if and only if $\eta = f^*(\Delta[E]) = 0$, so that the extensions in $\underline{\underline{V}}$ are classified by

$$\ker(f^* : H^2(G,A) \to H^2(F,A))$$

which by definition is $\tilde{V}(G,A)$.

III.4. The Category of Abelian Kernels

Let $\underline{\underline{V}}$ be a variety and let Q be in $\underline{\underline{V}}$. We have seen in Section 3 that a Q-module A can be the kernel of an extension $E : A \rightarrowtail G \twoheadrightarrow Q$ in $\underline{\underline{V}}$ only if $A \rtimes Q$ is in $\underline{\underline{V}}$. There is therefore considerable interest in the full subcategory $\underline{\underline{VMod}}_Q$ of $\underline{\underline{Mod}}_Q$ consisting of those Q-modules A with $A \rtimes Q$ in $\underline{\underline{V}}$. The category $\underline{\underline{VMod}}_Q$ is called the *category of abelian kernels*.

Let $f : G \to Q$ be a homomorphism (not necessarily surjective). Consider the diagram

(4.1)
$$\begin{array}{ccccc} & & & & G \\ & & & & \downarrow f \\ A & \stackrel{h}{\rightarrowtail} & A \rtimes Q & \stackrel{g}{\twoheadrightarrow} & Q \end{array}$$

By (II.2.3) (see also [43], Corollary VI.5.4) there is a one-to-one correspondence between maps $f' : G \to A \wr Q$ with $gf' = f$ and derivations $d : G \to A$, the correspondence being given by

(4.2) $$f'(x) = (dx, fx) \ , \ x \in F \ .$$

Now if $A \wr Q$ is in $\underline{\underline{V}}$, then every homomorphism $f' : G \to A \wr Q$ must vanish on VG , so that we may conclude from (4.2) that every derivation $d : G \to A$ vanishes on VG . We may thus state

PROPOSITION 4.1. *If* $A \wr Q$ *is in* $\underline{\underline{V}}$, *then for every group* G *and every homomorphism* $f : G \to Q$, *every derivation* $d : G \to A$ *vanishes on* VG .

Now consider the special case where $G = F_\infty$, the $\underline{\underline{Gr}}$-free group on generators $x_1, x_2, \ldots$. Let the variety be defined by the laws $v \in (v)$. If $A \wr Q$ is in $\underline{\underline{V}}$, then it follows by Proposition 4.1 that for every $f : F_\infty \to Q$ every derivation $d : F_\infty \to A$ vanishes on every $[v] \in F_\infty$, $v \in (v)$. Conversely, if for every $f : F_\infty \to Q$ every derivation $d : F_\infty \to A$ vanishes on all $[v] \in F_\infty$, $v \in (v)$, then $V(A \wr Q) = e$, so that $A \wr Q$ is in $\underline{\underline{V}}$. We have thus proved

PROPOSITION 4.2. *Let* $\underline{\underline{V}}$ *be defined by the laws* $v \in (v)$ *and let* Q *be in* $\underline{\underline{V}}$. *Then the* Q-*module* A *is in* $\underline{\underline{VMod}}_Q$ *if and only if for every* $f : F_\infty \to Q$ *every derivation* $d : F_\infty \to A$ *vanishes on* $[v] \in F_\infty$, $v \in (v)$.

Using Fox-derivatives (see Section II.2) we may rephrase Proposition 4.2 as follows.

COROLLARY 4.3. *Let* $\underline{\underline{V}}$ *be defined by the laws* $v \in (v)$ *and let* Q *be in* $\underline{\underline{V}}$. *Then the* Q-*module* A *is in* $\underline{\underline{VMod}}_Q$ *if and only if for every* $f : F_\infty \to Q$ *the elements* $f\partial_i[v] \in ZQ$, $v \in (v)$, $i = 1,2,\ldots$ *act as zero on* A .

PROOF: Suppose $f : F_\infty \to Q$ is given. Then we have a natural isomorphism (II.2.2)

$$(4.3) \qquad \eta_A : \mathrm{Der}(F_\infty, A) \xrightarrow{\sim} \mathrm{Hom}_{F_\infty}(IF_\infty, A)$$

associating with the derivation $d : F_\infty \to A$ the homomorphism $\psi : IF_\infty \to A$ defined by

$$(4.4) \qquad \psi(x-1) = dx \quad , \quad x \in F_\infty .$$

By Proposition II.2.2 we have

$$(4.5) \qquad dx = \psi(x-1) = \psi\Big(\sum_{i=1}^{\infty} \partial_i(x)(x_i-1)\Big) = \sum_{i=1}^{\infty} f\partial_i(x)\psi(x_i-1)$$

where we have used f to denote the map $ZF_\infty \to ZQ$ induced by $f : F_\infty \to Q$. Since IF_∞ is free on (x_i-1) , $i = 1,2,\ldots$ it follows that $dx = 0$ for all derivations $d : F_\infty \to A$ if and only if $f\partial_i(x)$, $i = 1,2,\ldots$ operate as zero on A .

Using Proposition 4.2 we thus conclude that A is in $\underline{\underline{VMod}}_Q$ if and only if for all $f : F_\infty \to Q$ the elements $f\partial_i[v]$, $i = 1,2,\ldots$, $v \in (v)$ operate as zero on A .

Denote by $I_V Q$ the 2-sided ideal of ZQ generated by all

$$(4.6) \qquad f\partial_i[v] \; , \; i = 1,2,\ldots \; , \; v \in (v) \; , \; f : F_\infty \to Q$$

and by $Z_V Q$ the quotient ring

$$(4.7) \qquad Z_V Q = ZQ/I_V Q .$$

We may then rephrase our result as follows.

COROLLARY 4.4. *The* Q-*module* A *is in* $\underline{\underline{VMod}}_Q$ *if and only if* A *is a* $Z_V Q$-*module*.

Next we give some examples:

(i) $\underline{\underline{V}} = \underline{\underline{Gr}}$. Of course, in this case, $\underline{\underline{VMod}}_Q = \underline{\underline{Mod}}_Q$.

(ii) $\underline{\underline{V}} = \underline{\underline{Ab}}$. Obviously every extension in $\underline{\underline{Ab}}$ is central, so that a module in $\underline{\underline{VMod}}_Q$ is necessarily a trivial Q-module. Conversely, if A is a trivial Q-module, then $A \rtimes Q = A \times Q$ is abelian, so that A is in $\underline{\underline{VMod}}_Q$. It follows that $\underline{\underline{VMod}}_Q$ consists precisely of the trivial Q-modules.

(iii) $\underline{\underline{V}} = \underline{\underline{B}}_q$. The law defining $\underline{\underline{B}}_q$ is x_1^q , so that we have to consider the elements

$$\partial_i[x_1^q] \in ZF_\infty \ , \ i = 1,2,... \tag{4.8}$$

It is easy to verify that

$$\partial_i[x_1^q] = \begin{cases} 1+x_1+x_1^2+...+x_1^{q-1} & , \ i = 1 \\ 0 & , \ i \neq 1 . \end{cases} \tag{4.9}$$

Now $f : F_\infty \to Q$ is allowed to send x_1 to an arbitrary element $a \in Q$. Thus we see that $I_V Q$ is the ideal of ZQ generated by all elements of the form

$$1+a+a^2+...+a^{q-1} \ , \ a \in Q . \tag{4.10}$$

(iv) $\underline{\underline{V}} = \underline{\underline{N}}_c$. We claim

PROPOSITION 4.5. *Let* $\underline{\underline{V}} = \underline{\underline{N}}_c$, *then* $Z_V Q = ZQ/(IQ)^c$.

PROOF: We proceed by induction on c . For c = 1 , example (ii)

establishes the result, for $\underline{\underline{N}}_1 = \underline{\underline{Ab}}$. Let $c \geq 2$. We first show that $I_V Q \subseteq IQ^c$. Let

(4.11) $$v = [x_1,[x_2,[x_3,\ldots,[x_c,x_{c+1}]\ldots]]]$$

be the law defining $\underline{\underline{V}} = \underline{\underline{N}}_c$. Set

(4.12) $$y = [x_2,[x_3,\ldots[x_c,x_{c+1}]\ldots]].$$

Then we have for any derivation $d : F_\infty \to ZF_\infty$

(4.13) $$\begin{aligned} d[x_1,y] &= dx_1+x_1dy - x_1yx_1^{-1}dx_1-x_1yx_1^{-1}y^{-1}dy \\ &= (1-x_1yx_1^{-1})dx_1 + x_1(1-yx_1^{-1}y^{-1})dy \end{aligned}$$

where we have used the fact that $d(z^{-1}) = -z^{-1}dz$, $z \in F_\infty$. Now let $f : F_\infty \to Q$ be any homomorphism. By induction $fdy \in IQ^{c-1}$, and by Lemma 4.6 below $f(1-x_1yx_1^{-1}) \in IQ^c$, so that indeed

(4.14) $$fd[x_1,y] \in IQ^c .$$

In particular we may set $d = \partial_i$, $i = 1,2,\ldots$ proving

(4.15) $$I_V Q \subseteq IQ^c .$$

Next we show that $IQ^c \subseteq I_V Q$. To do so we consider

(4.16) $$[v] = [x_1,[x_2,\ldots,[x_c,x_{c+1}]]] \in F_\infty ,$$

a homomorphism $f : F_\infty \to Q$ with $f(x_{c+1}) = e \in Q$, and the derivation

(4.17) $$\alpha\partial_{c+1} : F_\infty \to A$$

where $\alpha : ZF \to A$ is given by $\alpha(1) = a \in A$. We then have, using (4.13),

(4.18) $$\partial_{c+1}[v] = f(1-x_1yx_1^{-1})\partial_{c+1}x_1 + f(x_1(1-yx^{-1}y^{-1}))\partial_{c+1}y$$
$$= 0 + f(x_1-1)\partial_{c+1}y$$

since $\partial_{c+1}x_1 = 0$ and $f(y) = e$. By induction we obtain

(4.19) $$\partial_{c+1}[v] = f((x_1-1)(x_2-1)\ldots(x_c-1))\partial_{c+1}(x_{c+1})$$

so that

(4.20) $$\alpha\partial_{c+1}[v] = f((x_1-1)(x_2-1)\ldots(x_c-1))a \ .$$

Since $f(x_1),\ldots,f(x_c) \in Q$ may be chosen arbitrarily it follows that

(4.21) $$IQ^c \subseteq I_VQ \ .$$

Formulas (4.15) and (4.21) prove the proposition.

It remains to prove the

<u>LEMMA 4.6</u>. <u>If</u> $y \in F_c$, <u>then</u> $(1-y) \in IF^c$.

<u>PROOF</u>: We proceed by induction on c . For $c = 1$ the assertion is trivial. Let $c \geq 2$. We may obviously concentrate on elements of the form $y = [x,z]$, $x \in F$, $z \in F_{c-1}$. Then

$$\begin{aligned}1-y &= 1-xzx^{-1}z^{-1}\\ &= (zx-xz)x^{-1}z^{-1}\\ &= ((1-z)(1-x)-(1-x)(1-z))x^{-1}z^{-1} \ .\end{aligned}$$

Since we may assume inductively that $1-z \in IF^{c-1}$, it follows that $1-y \in IF^c$.

We conclude this section with the following general result.

PROPOSITION 4.7. The category of trivial Q-modules in $\underline{\underline{VMod}}_Q$ consists precisely of the groups in $\underline{\underline{V}} \cap \underline{\underline{Ab}}$.

PROOF: If A is in $\underline{\underline{VMod}}_Q$, then clearly A regarded as abelian group is in $\underline{\underline{V}} \cap \underline{\underline{Ab}}$ for it is an abelian subgroup of a group in $\underline{\underline{V}}$. Conversely, if A is in $\underline{\underline{V}} \cap \underline{\underline{Ab}}$ and the Q-module structure is trivial, then $A \rtimes Q = A \times Q$, which clearly lies in $\underline{\underline{V}}$. It follows that A is in $\underline{\underline{VMod}}_Q$.

Note that it follows from Proposition 4.7 that $\underline{\underline{V}}$ is of exponent q if and only if $Z_V Q$ is a Z/qZ-algebra. This may, of course, also be proved using formula (4.9).

III.5. The Coefficient Exact Sequences

In this section we shall deduce exact sequences for $\check{V}$ and V that correspond to the long exact (co)homology sequences in ordinary (co)-homology.

Let $\underline{\underline{V}}$ be any variety and let Q be in $\underline{\underline{V}}$.

PROPOSITION 5.1. The functor $Der(Q,-) : \underline{\underline{VMod}}_Q \to \underline{\underline{Ab}}$ is representable. In fact, there is a natural equivalence ζ with

$$(5.1) \qquad \zeta_A : Der(Q,A) \xrightarrow{\sim} Hom_{Z_V Q}(Z_V Q \otimes_Q IQ, A) .$$

PROOF: Using (II.2.2) and the fact that A is in $\underline{\underline{VMod}}_Q$ we have

$$\zeta_A : Der(Q,A) \overset{\eta_A}{\xrightarrow{\sim}} Hom_Q(IQ,A) \xrightarrow{\sim} Hom_{Z_V Q}(Z_V Q \otimes_Q IQ,A) .$$

COROLLARY 5.2. Let F be $\underline{\underline{V}}$-free. Then $Z_V F \otimes_F IF$ is a free $Z_V F$-module.

PROOF: The proof is the same as the proof of Proposition II.2.1. Of course, we have to use (5.1) and the fact that $A \rtimes F$ is in $\underline{\underline{V}}$. We will omit the details.

<u>THEOREM 5.3</u>. <u>Let</u> $A' \overset{\alpha'}{\rightarrowtail} A \overset{\alpha''}{\twoheadrightarrow} A''$ <u>be a short exact sequence in</u> $\underline{\underline{V}}\underline{\underline{Mod}}^{\ell}_{Q}$ <u>and let</u> $B' \overset{\beta'}{\rightarrowtail} B \overset{\beta''}{\twoheadrightarrow} B''$ <u>be a short exact sequence in</u> $\underline{\underline{V}}\underline{\underline{Mod}}^{r}_{Q}$. <u>Then there are exact sequences</u>

$$(5.2)\qquad \begin{array}{l} 0 \to \mathrm{Der}(Q,A') \xrightarrow{\alpha'_*} \mathrm{Der}(Q,A) \xrightarrow{\alpha''_*} \mathrm{Der}(Q,A'') \xrightarrow{\omega} \\ \xrightarrow{\omega} \tilde{V}(Q,A') \xrightarrow{\alpha'_*} \tilde{V}(Q,A) \xrightarrow{\alpha''_*} \tilde{V}(Q,A'') \end{array}$$

$$(5.3)\qquad \begin{array}{l} V(Q,B') \xrightarrow{\beta'_*} V(Q,B) \xrightarrow{\beta''_*} V(Q,B'') \xrightarrow{\sigma} \\ \xrightarrow{\sigma} B' \otimes_Q IQ \xrightarrow{\beta'_*} B \otimes_Q IQ \xrightarrow{\beta''_*} B'' \otimes_Q IQ \to 0 \end{array}$$

<u>PROOF</u>: Let $f : F \twoheadrightarrow Q$ be a $\underline{\underline{V}}$-free presentation of Q and consider the diagram

$$(5.4)\qquad \begin{array}{ccccccccc} & & & & \tilde{V}(Q,A') & \to & \tilde{V}(Q,A) & \to & \tilde{V}(Q,A'') \\ & & & \overset{\omega}{\nearrow} & \downarrow & & \downarrow & & \downarrow \\ \ldots \to \mathrm{Der}(Q,A) & \to & \mathrm{Der}(Q,A'') & \xrightarrow{\omega_Q} & H^2(Q,A') & \to & H^2(Q,A) & \to & H^2(Q,A'') \\ f^*\downarrow & & f^*\downarrow & & f^*\downarrow & & f^*\downarrow & & f^*\downarrow \\ \ldots \to \mathrm{Der}(F,A) & \to & \mathrm{Der}(F,A'') & \xrightarrow{\omega_F} & H^2(F,A') & \to & H^2(F,A) & \to & H^2(F,A'') \end{array}$$

We first note that for any Q-module M the map $f^* : \mathrm{Der}(Q,M) \to \mathrm{Der}(F,M)$ is monomorphic, so that the two left most vertical maps f^* are monomorphic. In order to prove the existence of ω and the exactness of the resulting sequence (5.2) it is enough to show that $\omega_F = 0$. By Proposition 5.1 we have

$$\mathrm{Der}(F,-) \cong \mathrm{Hom}_{Z_V Q}(Z_V F \otimes_F IF,-) \; : \; \underline{\underline{V}}\underline{\underline{Mod}}_Q \to \underline{\underline{Ab}}$$

and by Corollary 5.2 the module $Z_V F \otimes_F IF$ is $Z_V F$-free. It follows that $\mathrm{Der}(F,-) : \underline{\underline{V}}\underline{\underline{Mod}}_Q \to \underline{\underline{Ab}}$ is an exact functor; this implies that

$\omega_F = 0$.

The proof for the sequence (5.3) is dual; it may therefore be omitted.

Note that the analogous sequences associated with a short exact sequence in $\underline{\underline{Mod}}_Q$ (not $\underline{\underline{VMod}}_Q$) are not, in general, exact. Note also that despite the existence of the sequences (5.2), (5.3), the functors

$$\tilde{V}(Q,-) : \underline{\underline{VMod}}^{\ell}_Q \to \underline{\underline{Ab}} ,$$
$$V(Q,-) : \underline{\underline{VMod}}^{r}_Q \to \underline{\underline{Ab}}$$

are not, in general, derived functors in the module category $\underline{\underline{VMod}}_Q$.

III.6. The Coproduct Theorem for $\tilde{V}$ and V

Let G_i , $i = 1,2$ be groups in $\underline{\underline{V}}$ and let $G_1 *_V G_2$ be their varietal product (see I.3.4), i.e. their coproduct in the category $\underline{\underline{V}}$.

THEOREM 6.1. Let A be in $\underline{\underline{VMod}}^{\ell}_{G_V}$ and let B be in $\underline{\underline{VMod}}^{r}_{G_V}$. Then the coproduct injections induce isomorphisms

(6.1) $$\tilde{V}(G_1 *_V G_2, A) \xrightarrow{\sim} \tilde{V}(G_1, A) \oplus \tilde{V}(G_2, A) ,$$

(6.2) $$V(G_1, B) \oplus V(G_2, B) \xrightarrow{\sim} V(G_1 *_V G_2, B) .$$

PROOF: We only prove (6.1), the proof of (6.2) being dual. Let $f_i : F_i \to G_i$, $i = 1,2$, be $\underline{\underline{V}}$-free presentations. Consider the diagram

(6.3) $$\begin{array}{ccccc} V(F_1 * F_2) & \rightarrowtail & F_1 * F_2 & \xrightarrow{h} & F_1 *_V F_2 \\ \downarrow f' & & \downarrow f & & \downarrow f_V \\ V(G_1 * G_2) & \rightarrowtail & G_1 * G_2 & \xrightarrow{g} & G_1 *_V G_2 \end{array}$$

where the vertical maps are induced by f_1, f_2 . Note that f_V is a $\underline{\underline{V}}$-free presentation of $G_1 *_V G_2$. Set, for short,

$$(6.4)\qquad \begin{aligned} G &= G_1 * G_2 \quad , \quad G_V = G_1 *_V G_2 \, , \\ F &= F_1 * F_2 \quad , \quad F_V = F_1 *_V F_2 \, . \end{aligned}$$

Diagram (6.3) gives rise to a diagram of 5-term sequences

$$(6.5)\qquad \begin{array}{ccccccccc} 0 \to \mathrm{Der}(G_V,A) & \xrightarrow{g^*} & \mathrm{Der}(G,A) & \to & \mathrm{Hom}_{G_V}((VG)_{ab},A) & \xrightarrow{\delta^*} & H^2(G_V,A) & \xrightarrow{g^*} & H^2(G,A) \\ f_V^*\downarrow & & f^*\downarrow & & f'^*\downarrow & & f_V^*\downarrow & & f^*\downarrow \\ 0 \to \mathrm{Der}(F_V,A) & \xrightarrow{h^*} & \mathrm{Der}(F,A) & \to & \mathrm{Hom}_{F_V}((VF)_{ab},A) & \xrightarrow{\delta^*} & H^2(F_V,A) & \xrightarrow{h^*} & H^2(F,A) \end{array}$$

Since A is in $\underline{\underline{VMod}}_{G_V}$, it follows by Proposition 4.1 that every derivation $d : G \to A$ vanishes on VG, thus giving rise to a derivation $d' : G_V \to A$. Hence we get an isomorphism

$$g^* : \mathrm{Der}(G_V,A) \xrightarrow{\sim} \mathrm{Der}(G,A) \; .$$

Similarly, since A is in $\underline{\underline{VMod}}_{F_V}$, we have

$$h^* : \mathrm{Der}(F_V,A) \xrightarrow{\sim} \mathrm{Der}(F,A) \; .$$

This implies that both homomorphisms δ^* in (6.5) are monomorphic. Since f' in (6.3) is epimorphic, f'^* in (6.5) is monomorphic. By Corollary II.6.2 we have

$$\begin{aligned} H^2(G,A) &\xrightarrow{\sim} H^2(G_1,A) \oplus H^2(G_2,A) \; , \\ H^2(F,A) &\xrightarrow{\sim} H^2(F_1,A) \oplus H^2(F_2,A) \end{aligned}$$

so that

$$(6.6)\qquad \ker f^* = \ker f_1^* \oplus \ker f_2^* = \tilde{V}(G_1,A) \oplus \tilde{V}(G_2,A) \; .$$

Since $j_i : G_i \to G_V$ as well as $j_i' : F_i \to F_V$ have left inverses (see I.3.5) we conclude that

$$g^* : H^2(G_V,A) \to H^2(G,A) ,$$
$$h^* : H^2(F_V,A) \to H^2(F,A)$$

both split. Since f'^* is monomorphic, it follows that

$$\begin{aligned}\tilde{V}(G_V,A) &= \ker(f_V^* : H^2(G_V,A) \to H^2(F_V,A)) \\ &\cong \ker(f^* : H^2(G,A) \to H^2(F,A)) \\ &= \tilde{V}(G_1,A) \oplus \tilde{V}(G_2,A) ,\end{aligned}$$

the latter by (6.6). Since the isomorphism is clearly induced by the coproduct injections, this completes the proof.

III.7. The Change of Variety Exact Sequence

In this section we shall deduce an exact sequence that arises from considering the functors $V(-,-)$ and $W(-,-)$ for two varieties $\underline{\underline{V}},\underline{\underline{W}}$ with $\underline{\underline{V}} \subseteq \underline{\underline{W}}$. Thus let $\underline{\underline{V}} \subseteq \underline{\underline{W}}$, and let Q be in $\underline{\underline{W}}$. Suppose B is a Q/VQ-module.

PROPOSITION 7.1. *There is a natural transformation*

$$\tau_{Q,B} : W(Q,B) \to V(Q/VQ,B).$$

Moreover, $\tau_{Q,B}$ *is surjective if* B *is in* $\underline{\underline{VMod}}_{Q/VQ}$ *or if* Q *is in* $\underline{\underline{V}}$.

PROOF: Let $F \twoheadrightarrow Q$ be an $\underline{\underline{Gr}}$-free presentation of Q . Then $F/WF \twoheadrightarrow Q$ is a $\underline{\underline{W}}$-free presentation and $F/VF \twoheadrightarrow Q/VQ$ is a $\underline{\underline{V}}$-free presentation. Consider the diagram

$$
\begin{array}{ccccccc}
H_2(F/WF,B) & \to & H_2(Q,B) & \to & W(Q,B) & \to & 0 \\
\downarrow & & \downarrow & & \downarrow{\scriptstyle \tau_{Q,B}} & & \\
H_2(F/VF,B) & \to & H_2(Q/VQ,B) & \to & V(Q/VQ,B) & \to & 0 \\
\downarrow & & \downarrow & & & & \\
B \otimes_F (VF/WF)_{ab} & \xrightarrow{\alpha} & B \otimes_Q (VQ)_{ab} & & & & \\
\downarrow & & \downarrow & & & & \\
B \otimes_F I(F/WF) & & B \otimes_Q IQ & & & & \\
\beta\downarrow & & \gamma\downarrow & & & & \\
B \otimes_F I(F/VF) & & B \otimes_Q I(Q/VQ) & & & & \\
\downarrow & & \downarrow & & & & \\
0 & & 0 & & & &
\end{array}
\tag{7.1}
$$

Plainly $\tau_{Q,B}$ yields a natural transformation. If B is in $\underline{\underline{VMod}}_{Q/VQ}$ then by Lemma 7.2 β as well as γ are isomorphisms. Since α is surjective it easily follows that $\tau_{Q,B}$ is surjective, also. If Q is in $\underline{\underline{V}}$, then $VQ = e$, and it is obvious that $\tau_{Q,B}$ is surjective, since it is induced by the identity of $H_2(Q,B)$.

It remains to prove the following

<u>LEMMA 7.2</u>. <u>Let</u> B <u>be in</u> $\underline{\underline{VMod}}^r_{Q/VQ}$. <u>Then</u>

$$B \otimes_Q IQ \cong B \otimes_Q I(Q/VQ) .$$

<u>PROOF</u>: For any A in $\underline{\underline{VMod}}^{\ell}_{Q/VQ}$ Proposition 4.1 yields

$$\mathrm{Der}(Q,A) \cong \mathrm{Der}(Q/VQ,A) ,$$

natural in A . Using (II.2.2) we obtain

$$\mathrm{Hom}_Q(IQ,A) \cong \mathrm{Hom}_Q(I(Q/VQ),A) .$$

Since A is in $\underline{\underline{VMod}}_{Q/VQ}$ we get from this

$$\mathrm{Hom}_Q(Z_V(Q/VQ)\otimes_Q IQ,A) \cong \mathrm{Hom}_Q(Z_V(Q/VQ)\otimes_Q I(Q/VQ),A)$$

again natural in A. We may thus conclude

$$Z_V(Q/VQ) \otimes_Q IQ \cong Z_V(Q/VQ) \otimes_Q I(Q/VQ) .$$

Hence for any B in $\underline{\underline{VMod}}^r_{Q/VQ}$ we have

$$B \otimes_Q IQ = B \otimes_Q Z_V(Q/VQ) \otimes_Q IQ \cong$$

$$\cong B \otimes_Q Z_V(Q/VQ) \otimes_Q I(Q/VQ) \cong B \otimes_Q I(Q/VQ) ,$$

thus completing the proof of Lemma 7.2.

COROLLARY 7.3. (Leedham-Green [55]). *Let* $\underline{\underline{V}} \subseteq \underline{\underline{W}}$ *and let* Q *be in* $\underline{\underline{V}}$. *Suppose* B *is in* $\underline{\underline{VMod}}_Q$. *If* $F/VF \twoheadrightarrow Q$ *is a* $\underline{\underline{V}}$*-free presentation, then there is an exact sequence*

$$(7.2) \qquad W(F/VF,B) \to W(Q,B) \xrightarrow{\tau_{Q,B}} V(Q,B) \to 0$$

PROOF: This can be read off from diagram (7.1). For under the hypotheses stated the map β is isomorphic and $Q = Q/VQ$. By definition of $W(F/VF,B)$ we have

$$(7.3) \qquad W(F/VF,B) \cong B \otimes_F (VF/WF)_{ab} .$$

An application of the ker - coker sequence then yields the result.

REMARK: Let Q be an arbitrary group in $\underline{\underline{W}} = \underline{\underline{Gr}}$ and let $\underline{\underline{V}} = \underline{\underline{Ab}}$. Then Proposition 7.1 yields a surjective map

$$(7.4) \qquad \tau_{Q,B} : H_2(Q,B) \twoheadrightarrow \mathrm{Tor}(Q_{ab},B)$$

for B in $\underline{\underline{VMod}}_{Q_{ab}}$, i.e. for a trivial Q-module. Obviously it may be identified with the map Σ' in the universal coefficient exact sequence (II.5.2) for $n = 2$. If Q is in $\underline{\underline{V}}$, i.e. if Q is abelian we may give an example for (7.2); for we have

$$W(F/VF,B) = H_2(F_{ab},B) = B \otimes H_2(F_{ab}) .$$

This obviously maps onto $B \otimes H_2(Q)$, the kernel of Σ' in the universal coefficient exact sequence.

We finally note that one obviously gets results dual to Proposition 7.1 and Corollary 7.2 by considering the functors $\tilde{W}$ and $\tilde{V}$. We leave the details to the reader.

III.8. The Universal Coefficient Exact Sequences in a Variety

In this section we deduce exact sequences for $\tilde{V}$, V that correspond to the universal coefficient sequences (II.5.1), (II.5.2) for $n = 2$.

THEOREM 8.1. Let $\underline{V}$ be a variety of exponent q and let Q be in $\underline{V}$. Suppose C is in $\underline{V} \cap \underline{\underline{Ab}}$. Then there are exact sequences

$$(8.1) \qquad 0 \to \operatorname{Ext}^1_{Z/qZ}(Q^q_{ab},C) \xrightarrow{\Sigma} \tilde{V}(Q,C) \xrightarrow{\Pi} \operatorname{Hom}(V^qQ,C) \to \operatorname{Ext}^2_{Z/qZ}(Q^q_{ab},C) \to \dots$$

$$(8.2) \qquad \dots \to \operatorname{Tor}_2^{Z/qZ}(Q^q_{ab},C) \to V^qQ \otimes C \xrightarrow{\Pi'} V(Q,C) \xrightarrow{\Sigma'} \operatorname{Tor}_1^{Z/qZ}(Q^q_{ab},C) \to 0$$

PROOF: We only prove (8.2); the proof of (8.1) being dual. First we recall from Proposition 4.7 that if C is in $\underline{V} \cap \underline{\underline{Ab}}$, then C is in $\underline{\underline{VMod}}_Q$. Next we consider a $\underline{V}$-free presentation

$$R \overset{h}{\rightarrowtail} F \overset{g}{\twoheadrightarrow} Q .$$

Then we have an exact sequence

$$(8.3) \qquad 0 \to V^qQ \to R/F \#_q R \xrightarrow{h_*} F^q_{ab} \xrightarrow{g_*} Q^q_{ab} \to 0 .$$

Denote by K the image of h_* and consider the two resulting short exact sequences

(8.4) $$0 \to V^qQ \to R/F \#_q R \to K \to 0 ,$$

(8.5) $$0 \to K \to F^q_{ab} \to Q^q_{ab} \to 0 .$$

Tensoring with C yields the diagram

(8.6)
$$\begin{array}{ccccccc}
 & & & & 0 & & 0 \\
 & & & & \downarrow & & \downarrow \\
 & & & \nearrow & V(Q,C) & \dashrightarrow & \mathrm{Tor}_1^{Z/qZ}(Q^q_{ab},C) \\
 & & & & \downarrow & & \downarrow \\
\ldots \to \mathrm{Tor}_1^{Z/qZ}(K,C) & \to & V^qQ \otimes C & \to & R/F \#_q R \otimes C & \twoheadrightarrow & K \otimes C \\
 & & & & \downarrow & & \downarrow \\
 & & & & F^q_{ab} \otimes C & = & F^q_{ab} \otimes C \\
 & & & & \downarrow & & \downarrow \\
 & & & & Q^q_{ab} \otimes C & = & Q^q_{ab} \otimes C \\
 & & & & \downarrow & & \downarrow \\
 & & & & 0 & & 0
\end{array}$$

Since F^q_{ab} is Z/qZ-free, it follows from the long exact Tor-sequence that

$$\mathrm{Tor}_1^{Z/qZ}(K,C) \cong \mathrm{Tor}_2^{Z/qZ}(Q^q_{ab},C) ,$$

thus completing the proof.

We remark that sequence (8.2) can be continued to the left (and similarly sequence (8.1) to the right) using the long exact sequence associated with (8.4). We obtain, $n \geqslant 1$

(8.7) $$\ldots \to \mathrm{Tor}_n^{Z/qZ}(V^qQ,C) \to \mathrm{Tor}_n^{Z/qZ}(R/F \#_q R,C) \to$$
$$\to \mathrm{Tor}_{n+1}^{Z/qZ}(Q^q_{ab},C) \to \mathrm{Tor}_{n-1}^{Z/qZ}(V^qQ,C) \to \ldots$$

and the analogous Ext-sequence.

COROLLARY 8.2. *If* $\underline{\underline{V}}$ *is of exponent* $q = 0$, *then we have split short exact sequences*

$$(8.8) \qquad 0 \to \operatorname{Ext}^1_Z(Q_{ab},C) \to \tilde{V}(Q,C) \to \operatorname{Hom}(VQ,C) \to 0 ,$$

$$(8.9) \qquad 0 \to VQ \otimes C \to V(Q,C) \to \operatorname{Tor}^Z_1(Q_{ab},C) \to 0 .$$

PROOF: Again we only prove the homology part. Since $\operatorname{Tor}^Z_2 = 0$, it is enough to prove that sequence (8.9) splits. But if $q = 0$, then F_{ab} is free abelian. Hence K is free abelian and (8.4) splits. It is then obvious from diagram (8.6) that this yields a splitting of (8.9).

For varieties $\underline{\underline{V}}$ of exponent $q \neq 0$, we are, of course, interested in the case where

$$(8.10) \qquad \begin{aligned} \operatorname{Tor}^{Z/qZ}_2(Q^q_{ab},C) &\cong \operatorname{Tor}^{Z/qZ}_1(K,C) = 0 , \\ \operatorname{Ext}^2_{Z/qZ}(Q^q_{ab},C) &\cong \operatorname{Ext}^1_{Z/qZ}(K,C) = 0 . \end{aligned}$$

This is so, if, for example, K is projective. But by Lemma II.5.2 K is projective if and only if Q^q_{ab} is. We thus obtain

COROLLARY 8.2. *If* $\underline{\underline{V}}$ *is of exponent* $q > 0$, *and if* Q^q_{ab} *is* Z/qZ-*projective, then for* C *in* $\underline{\underline{V}} \cap \underline{\underline{Ab}}$

$$(8.11) \qquad \tilde{V}(Q,C) \cong \operatorname{Hom}(V^qQ,C) ,$$

$$(8.12) \qquad V(Q,C) \cong V^qQ \otimes C .$$

We finally note that the statement analogous to Proposition II.5.4 is also true: Let C be in $\underline{\underline{V}} \cap \underline{\underline{Ab}}$ and let $E : C \rightarrowtail G \twoheadrightarrow Q$ be an extension in $\underline{\underline{V}}$ characterized by $\Delta[E] = \xi \in \tilde{V}(Q,C)$. Then

$$(8.13) \qquad \Pi\Delta[E] : V^qQ \to C$$

is the homomorphism δ_*^E in the 5-term sequence

$$V^qG \to V^qQ \xrightarrow{\delta_*^E} C \to G^q_{ab} \to Q^q_{ab} \to 0 \ .$$

Since the proof of this fact is analogous to that of Proposition II.5.4 we leave it to the reader.

CHAPTER IV

THE LOWER CENTRAL SERIES

The key result of this Chapter is Theorem 1.1. It enables one to conclude that a group homomorphism $f : K \to G$ satisfying certain homological hypotheses induces isomorphisms between the quotients of K and G by the terms of the lower central series. The applications of this simple theorem are surprisingly numerous; the presentation of some of them takes up the rest of this chapter. After some preparation in Sections 1, 2 we are able to prove (Section 3) most of the known theorems giving sufficient conditions for a subset of a $\underline{\underline{V}}$-free group to generate a $\underline{\underline{V}}$ -free subgroup. In Section 4 we obtain some of the better known theorems on splitting groups in a variety, in particular the famous theorem of P. Hall [38] about splitting groups in a nilpotent variety. In Section 5 we apply an extension of our basic Theorem 1.1 to para-free groups and prove and extend some known results. Section 6 deals with the notion of deficiency of groups. Here our basic theorem yields some interesting results about the deficiency of nilpotent groups. In Section 7 we prove a theorem on groups given by a special presentation that has as a corollary the famous theorem of Magnus [60] that a group given by $n+r$ generators and r relators which may also be generated by n elements is free. In Section 8 we turn to finite groups to prove the Huppert-Thompson-Tate-theorem on the existence of normal p-complements using our Theorem 1.1.

It is a most striking fact that although the statements of many of the theorems of this chapter do not require any homological terminology, their proof nevertheless uses homological machinery. Here we see the homology theory of groups at its best: as a tool, to prove theorems of a non-homological nature.

Theorem 1.1 for $\underline{\underline{V}} = \underline{\underline{Gr}}$ is contained in Stallings [74], Stammbach [75], where some applications are to be found, also. Much of the rest is in Stammbach [76], [77], [78], [79]. Tate's paper [82] is related to Section 8. Our information about varieties of groups used in this chapter stems from H. Neumann [64]; most of the results on parafree groups we present in Section 5 come from Baumslag [13], [14].

IV.1. The Basic Theorem

Let $\underline{\underline{V}}$ be a variety of groups, and let G be in $\underline{\underline{V}}$. For q a non-negative integer we shall consider the groups

$$V(G,\mathbb{Z}/q\mathbb{Z}) = V^q G \tag{1.1}$$

as defined in Sections III.1 , III.2. Recall that for $\underline{\underline{V}} = \underline{\underline{Gr}}$ we have $V^q G = H_2^q G$. We shall also consider the lower central (q) series of G

$$G_1^q = G \ , \ G_{i+1}^q = G \#_q G_i^q \ , \qquad i = 1,2,\ldots \tag{1.2}$$

as defined in (I.1.2). We shall set

$$G_\omega^q = \bigcap_{i=1}^{\infty} G_i^q \ , \ G_\omega = \bigcap_{i=1}^{\infty} G_i \ . \tag{1.3}$$

The basic theorem of which we shall make numerous applications in this chapter is as follows.

THEOREM 1.1. *Let $f : K \to G$ be a homomorphism of groups in $\underline{\underline{V}}$. Suppose that f induces an isomorphism $f_* : K_{ab}^q \tilde{\to} G_{ab}^q$ and an epimorphism $f_* : V^q K \twoheadrightarrow V^q G$. Then the map f induces an isomorphism*

$$f_i^q : K/K_i^q \tilde{\to} G/G_i^q \tag{1.4}$$

for every $i \geq 1$, and a monomorphism $f_\omega^q : K/K_\omega^q \rightarrowtail G/G_\omega^q$.

<u>PROOF</u>: We proceed by induction. For $i = 1$ the conclusion is trivial, and for $i = 2$ it is part of the hypotheses. For $i > 2$ we consider the exact sequences (III.2.7)

(1.6)
$$\begin{array}{ccccccccc} V^qK & \to & V^q(K/K^q_{i-1}) & \to & K^q_{i-1}/K^q_i & \to & K^q_{ab} & \to & (K/K^q_{i-1})^q_{ab} \to 0 \\ \alpha_5\downarrow & & \alpha_4\downarrow & & \alpha_3\downarrow & & \alpha_2\downarrow & & \alpha_1\downarrow \\ V^qG & \to & V^q(G/G^q_{i-1}) & \to & G^q_{i-1}/G^q_i & \to & G^q_{ab} & \to & (G/G^q_{i-1})^q_{ab} \to 0 \end{array}$$

and the map induced by $f : K \to G$. By induction α_1, α_4 are isomorphisms. By hypothesis α_2 is an isomorphism, and α_5 is an epimorphism. By the 5-lemma this implies that α_3 is an isomorphism. Next we apply the 5-lemma to the diagram

(1.7)
$$\begin{array}{ccccc} K^q_{i-1}/K^q_i & \rightarrowtail & K/K^q_i & \twoheadrightarrow & K/K^q_{i-1} \\ \alpha_3\downarrow & & \downarrow f^q_i & & \downarrow f^q_{i-1} \\ G^q_{i-1}/G^q_i & \rightarrowtail & G/G^q_i & \twoheadrightarrow & G/G^q_{i-1} \end{array}$$

By the above α_3 is isomorphic; by induction f^q_{i-1} is isomorphic. Hence f^q_i is isomorphic, and the first part of the conclusion is proved.

In order to prove the statement about f_ω consider the diagram

(1.8)
$$\begin{array}{ccc} K/K^q_\omega & \xrightarrow{f^q_\omega} & G/G^q_\omega \\ \downarrow & & \downarrow \\ K/K^q_i & \xrightarrow{f^q_i} & G/G^q_i \end{array}$$

and suppose $xK^q_\omega \in \ker f^q_\omega$. Then, for all $i \geq 1$, we have $f^q_i(xK^q_i) = G^q_i$. But, since f^q_i is an isomorphism by (1.4), it follows that $x \in K^q_i$, $i \geq 1$. Hence $x \in K^q_\omega$, and f^q_ω is monomorphic.

Under the hypothese of Theorem 1.1 the map f^q_ω is not, in general, epimorphic. A counterexample is as follows (see [74]). Let

$K = G = F(x,y)$, the $\underline{\underline{Gr}}$-free group on two generators x,y . Consider $f : K \to G$ given by $f(x) = x$, $f(y) = y[x,y]$. Taking $q = 0$, it is immediate that $f_* : K_{ab} \to G_{ab}$ is an isomorphism. Also, $f_* : H_2K \to H_2G(= 0)$ is an epimorphism. Since absolutely free groups are residually nilpotent, we have $K_\omega = G_\omega = e$, so that $f_\omega = f$. But of course, f is not surjective, for $y \notin \operatorname{im} f$.

We remark however that quite generally, if f is epimorphic, f^q_ω also is. It follows, that whenever the map $f : K \to G$ in Theorem 1.1 is epimorphic, we may conclude that f^q_ω is an isomorphism.

COROLLARY 1.2. *Let* K,G *be groups that are nilpotent* (q). *Suppose that* $f : K \to G$ *induces an isomorphism* $f_* : K^q_{ab} \xrightarrow{\sim} G^q_{ab}$ *and an epimorphism* $f_* : V^qK \twoheadrightarrow V^qG$. *Then* f *itself is an isomorphism.*

PROOF: By definition (see Section I.1) a group G is called nilpotent (q) if there exists $n \geq 1$ with $G^q_n = e$. The result then easily follows from Theorem 1.1.

COROLLARY 1.3. *Let* G *be a group in* $\underline{\underline{V}}$, *with* $V^qG = 0$. *In case* q *is not a prime, suppose that* G^q_{ab} *is free in* $\underline{\underline{V}} \cap \underline{\underline{Ab}}$. *Then there exists a* $\underline{\underline{V}}$*-free group* F *and a homomorphism* $f : F \to G$ *such that the map* f *induces an isomorphism*

$$f^q_i : F/F^q_i \xrightarrow{\sim} G/G^q_i ,$$

for every $i \geq 1$, *and a monomorphism* $f^q_\omega : F/F^q_\omega \rightarrowtail G/G^q_\omega$.

PROOF: Let (x_j) , $j \in J$ be a set of elements in G , whose images in G^q_{ab} form a basis. Consider the $\underline{\underline{V}}$-free group F on the set (y_j) , $j \in J$ and the map $f : F \to G$ defined by $f(y_j) = x_j$, $j \in J$. Obviously the hypotheses of Theorem 1.1 are satisfied, so that the conclusion follows.

COROLLARY 1.4. *Let* $\underline{\underline{V}}$ *be a nilpotent variety of exponent* $q = 0$. *If* G *is a finitely generated group in* $\underline{\underline{V}}$ *with* $V^pG = 0$ *for all primes* p , *then* G *is* $\underline{\underline{V}}$*-free*.

PROOF: Since G is finitely generated nilpotent it is finitely related. Hence H_2G is finitely generated and, a fortiori, VG is finitely generated. Since by (III.8.9)

$$V^qG \cong VG \otimes Z/qZ \oplus Tor(G_{ab}, Z/qZ)$$

it follows, that G_{ab} is torsionfree, hence free, and that $VG = 0$. The conclusion is then obtained by applying Corollaries 1.3 and 1.2.

REMARK. It is easy to see that the group G in the nilpotent variety $\underline{\underline{V}}$ of exponent $q = 0$ is $\underline{\underline{V}}$-free if and only if $\tilde{V}(G,A) = 0$ for all abelian groups A . By (III.8.8) we have

$$\tilde{V}(G,A) \cong Ext(G_{ab},A) \oplus Hom(VG,A)$$

Now if $\tilde{V}(G,A) = 0$, for all A , then G_{ab} must be free abelian and $VG = 0$. An application of Corollaries 1.3 and 1.2 then shows that G is free. Conversely, if G is free then $\tilde{V}(G,A) = 0$ by (III.1.11).

IV.2. Free Subgroup Theorems

In this section we shall state some results on the existence of $\underline{\underline{V}}$-free subgroups of a group G in $\underline{\underline{V}}$.

We begin with some preliminaries. Let p denote a prime or zero. We shall say that *the variety* $\underline{\underline{V}}$ *satisfies property* (P_p) if the $\underline{\underline{V}}$-free groups are residually nilpotent (p).

Recall from Section I.1 that nilpotent (O) means just nilpotent and that for $p \neq 0$, nilpotent (p) means nilpotent and of finite p exponent (see Lemma I.1.1).

LEMMA 2.1. A group G is residually nilpotent (p) if and only if $G_\omega^p = e$.

PROOF: Suppose G is residually nilpotent (p). Let $e \neq x \in G$. Then there exists $N \triangleleft G$ with $x \notin N$ and G/N nilpotent (p) of class c , say. It is then clear that $G_{c+1}^p \subseteq N$ so that $x \notin G_{c+1}^p$. It follows that $G_\omega^p = e$. Conversely, let $G_\omega^p = e$. Then given $e \neq x \in G$ there exists an integer n such that $x \notin G_n^p$. Since G/G_n^p is nilpotent (p) we may conclude that G is residually nilpotent (p).

Note that if G is residually a finite p-group, then G is residually nilpotent (p). We continue with some examples of varieties satisfying (P_p) :

(i) $\underline{\underline{V}} = \underline{\underline{Gr}}$; the absolutely free groups are residually nilpotent, and residually finite p-groups.

(ii) $\underline{\underline{V}} = \underline{\underline{N}}_c$; the free nilpotent groups are residually finite p-groups [34].

(iii) $\underline{\underline{V}} = \underline{\underline{S}}_\ell$; the free soluble groups are residually nilpotent and residually finite p-groups [34].

(iv) $V = \underline{\underline{P}}_{(c_1,c_2,\ldots,c_k)}$; the free groups in any polynilpotent variety are residually nilpotent and residually finite p-groups [34].

Note that example (iv) contains examples (ii) and (iii). We may now state the following general theorem.

THEOREM 2.2. (i) Let $\underline{\underline{V}}$ be a variety satisfying (P_o) and let G be a group in $\underline{\underline{V}}$ with G_{ab} free in $\underline{\underline{V}} \cap \underline{\underline{Ab}}$ and $VG = 0$. Let (x_j) , $j \in J$ be a set of elements in G whose images in G_{ab} freely (in $\underline{\underline{V}} \cap \underline{\underline{Ab}}$) generate a direct summand. Then (x_j) , $j \in J$ freely generates a $\underline{\underline{V}}$-free subgroup F of G .

(ii) *Let* $\underline{\underline{V}}$ *be a variety satisfying* (P_p) *for a prime* p *and let* G *be a group in* $\underline{\underline{V}}$ *with* $V^pG = 0$. *Let* (x_j) , $j \in J$ *be a set of elements in* G *whose images in* G^p_{ab} *are linearly independent. Then* (x_j), $j \in J$ *freely generates a* $\underline{\underline{V}}$*-free subgroup* F *of* G .

PROOF: In both cases we may enlarge the set (x_j) , $j \in J$ to a set of elements of G whose images in G^p_{ab} ($p = 0$ in case (i), p the given prime in case (ii)) form a basis. We shall prove that this larger set (which we again call (x_j) , $j \in J$) freely generates a $\underline{\underline{V}}$-free subgroup F of G . To do so, take F to be the $\underline{\underline{V}}$-free group on (y_j) , $j \in J$ and define $f : F \to G$ by $f(y_j) = x_j$. Clearly f satisfies the hypotheses of Theorem 1.1 whence we obtain a monomorphism $f^p_\omega : F/F^p_\omega \rightarrowtail G/G^p_\omega$. Since $\underline{\underline{V}}$ satisfies (P_p) we have $F^p_\omega = e$, and f^p_ω factors as

$$F \xrightarrow{f} G \to G/G^p_\omega .$$

It follows that f itself is monomorphic and the proof is complete.

The following example shows that it is not enough to suppose in Theorem 2.2 (i) that the images of (x_j) , $j \in J$ freely (in $\underline{\underline{V}} \cap \underline{\underline{Ab}}$) generate a free subgroup in G_{ab} . Take G to be the $\underline{\underline{V}}$-free group in two generators a,b where $\underline{\underline{V}}$ is the variety given by the laws $[x[y,z]]$ and $[x,y]^2$. It is obvious that G_{ab} is free abelian, so that the images of a^2,b in G_{ab} generate a free subgroup. However they do not generate a $\underline{\underline{V}}$-free subgroup of G because

$$\begin{aligned}[a^2,b] &= a(aba^{-1}b^{-1})a^{-1}(aba^{-1}b^{-1}) \\ &= [a,[a,b]][a,b]^2 \\ &= e .\end{aligned}$$

Hence a^2,b generate an abelian subgroup.

We next note that if G is a $\underline{\underline{V}}$-free group in a variety $\underline{\underline{V}}$ satisfying property (P_p) then G_{ab} is either free abelian or a p-group. Consequently if $\underline{\underline{V}}$ satisfies property (P_p) for more than one prime, it must be a variety of exponent zero. This remark will be used in the proof of

COROLLARY 2.3. *Let $\underline{\underline{V}}$ be a variety satisfying (P_p) for infinitely many primes. Let G be a group in $\underline{\underline{V}}$ with G_{ab} free abelian and $VG = 0$. Let (x_j), $j \in J$ be a set of elements in G whose images in G_{ab} freely generate a free subgroup. Then (x_j), $j \in J$ freely generates a $\underline{\underline{V}}$-free subgroup of G.*

PROOF: By Proposition I.3.1 it is enough to prove that every finite subset of (x_j), $j \in J$ generates a $\underline{\underline{V}}$-free subgroup. But for a finite subset (x_j), $j \in J'$ there exists a prime p such that $\underline{\underline{V}}$ satisfies (P_p) and such that the images of (x_j), $j \in J'$ in G^p_{ab} are linearly independent. The result then follows from Theorem 2.2 (ii), if we can show that $V^pG = 0$. To see this we recall that by the remark just above Corollary 2.3 the variety $\underline{\underline{V}}$ must be of exponent zero, so that we may apply Corollary III.8.2. We then have

$$V^pG \cong VG \otimes Z/pZ \oplus \mathrm{Tor}(G_{ab}, Z/pZ)$$

which clearly must be zero. Thus the proof is complete.

IV.3. Subgroups of $\underline{\underline{V}}$-free Groups

In this section we shall apply the results of Section 2 to give sufficient conditions for a subgroup of a $\underline{\underline{V}}$-free group to be $\underline{\underline{V}}$-free.

We first recall that by Schreier's theorem every subgroup of a $\underline{\underline{Gr}}$-free group is $\underline{\underline{Gr}}$-free. Also, obviously the varieties $\underline{\underline{Ab}}$ and $\underline{\underline{Ab}}_p$ for any prime p have that property. It is known (see [64]) that these are the only varieties of that kind (Schreier varieties). On the other hand there are many results stating sufficient conditions for a subset of a $\underline{\underline{V}}$-free group to generate a $\underline{\underline{V}}$-free subgroup. Most of these results concern varieties $\underline{\underline{V}}$ which satisfy property (P_p) for at least one prime p or $p = 0$. We shall get practically all of these results as direct corollaries of Theorem 2.2.

THEOREM 3.1. (i) (Hall [38], Mostowski [63]) *Let* $\underline{\underline{V}}$ *be a variety satisfying* (P_o). *Let* F *be a* $\underline{\underline{V}}$-*free group and let* (x_j), $j \in J$ *be a set of elements in* F *whose images in* F_{ab} *freely* (in $\underline{\underline{V}} \cap \underline{\underline{Ab}}$) *generate a direct summand. Then* (x_j), $j \in J$ *freely generates a* $\underline{\underline{V}}$-*free subgroup of* F.

(ii) *Let* $\underline{\underline{V}}$ *be a variety satisfying* (P_p) *for a certain prime* p. *Let* F *be a* $\underline{\underline{V}}$-*free group and let* (x_j), $j \in J$ *be a set of elements in* F *whose images in* F^p_{ab} *are linearly independent. Then* (x_j), $j \in J$ *freely generates a* $\underline{\underline{V}}$-*free subgroup of* F.

PROOF: The proof is immediate from Theorem 2.2 since for a $\underline{\underline{V}}$-free group F the group F_{ab} is free in $\underline{\underline{V}} \cap \underline{\underline{Ab}}$ and $V^pF = 0$ for all primes p and for $p = 0$.

COROLLARY 3.2. (Baumslag [12]) *Let* $\underline{\underline{V}}$ *be a variety satisfying* (P_p) *for infinitely many primes. Let* F *be a* $\underline{\underline{V}}$-*free group and let* (x_j), $j \in J$ *be a set of elements in* F *whose images in* F_{ab} *freely generate a free subgroup. Then* (x_j), $j \in J$ *freely generates a* $\underline{\underline{V}}$-*free subgroup of* F.

PROOF: The proof is immediate from Corollary 2.3.

PROPOSITION 3.3. (P.M. Neumann [65]) *Let* $\underline{\underline{V}}$ *be a variety whose free groups are residually nilpotent* π-*groups for some fixed non empty set*

π of primes. Let F be a $\underline{\underline{V}}$-free group and let (x_j) , $j \in J$ be a set of elements in F whose images in F_{ab} are linearly independent. Suppose that $F_{ab}/\langle x_j \rangle$ does not contain π-torsion. Then (x_j) , $j \in J$ freely generates a $\underline{\underline{V}}$-free subgroup of F .

PROOF: We start with the following remark. It is clear that F_{ab} is residually an (abelian) π-group. Thus F_{ab} is either free abelian or a free Z/qZ-module with q being a product of primes in π . In the latter case it is easy to see that F is even a residually nilpotent $\bar{\pi}$-group where $\bar{\pi}$ is the set of primes dividing q . In the sequel we shall therefore replace π by $\bar{\pi}$ in this case.

Denote by W the subgroup of F_{ab} generated by the images of (x_j) , $j \in J$. Consider the exact sequence

$$(3.1) \qquad W \rightarrowtail F_{ab} \twoheadrightarrow F_{ab}/\langle x_j \rangle .$$

Tensoring (3.1) with Z/pZ where $p \in \pi$ we obtain the exact sequence

$$(3.2) \qquad W \otimes Z/pZ \rightarrowtail F^p_{ab} \twoheadrightarrow F^p_{ab}/\langle x_j \rangle$$

since by hypothesis $F_{ab}/\langle x_j \rangle$ does not contain p-torsion. Since the images of (x_j) , $j \in J$ in F_{ab} are linearly independent their images in $W \otimes Z/pZ$ must form a basis. Proceeding as in the proof of Theorem 2.2 (ii) we thus obtain a map $f : F' \to F$, where F' is $\underline{\underline{V}}$-free on (y_j) , $j \in J$ and $f(y_j) = x_j$, with the property that for $p \in \pi$ the induced map

$$(3.3) \qquad f^p_\omega : F'/F'^p_\omega \to F/F^p_\omega$$

is monomorphic. This way we finally obtain a map

$$(3.4) \qquad f^*_\omega : F'/\bigcap_{p\in\pi} F'^p_\omega \to F/\bigcap_{p\in\pi} F^p_\omega$$

which obviously is monomorphic, also. It remains to prove that for a

$\underline{\underline{V}}$-free group F

(3.5) $$\bigcap_{p\in\pi} F_{\omega}^{p} = e \ .$$

Thus let $e \neq x \in F$. Then there exists $N_1 \triangleleft F$ with $x \notin N_1$ and F/N_1 finitely generated $\underline{\underline{V}}$-free. Also, there exists $N_2 \triangleleft F$ with $N_1 \subseteq N_2$, $x \notin N_2$ and F/N_2 a finitely generated, hence finite, nilpotent π-group. Finally there exists $p \in \pi$ and $N_3 \triangleleft F$ with $N_2 \subseteq N_3$ $x \notin N_3$ and F/N_3 a finite p-group. Since finite p-groups are nilpotent (p), there exists $m \geq 1$ with $F_{m+1}^{p} \subseteq N_3$, so that certainly

(3.6) $$x \notin F_{m+1}^{p}$$

and, a fortiori, $x \notin \bigcap_{p\in\pi} F_{m+1}^{p}$. Thus the proof is complete.

IV.4. Splitting Groups in $\underline{\underline{V}}$

A group G in $\underline{\underline{V}}$ is called a <u>splitting group</u> in $\underline{\underline{V}}$ if to every surjective map $f : K \twoheadrightarrow G$ there exists $t : G \to K$ with $ft = 1_G$.

A group G in $\underline{\underline{V}}$ is called a <u>retract of the $\underline{\underline{V}}$-free group</u> F if there is a $\underline{\underline{V}}$-presentation $h : F \twoheadrightarrow G$ such that there exists $s : G \to F$ with $hs = 1_G$.

<u>LEMMA 4.1</u>. <u>Let</u> G <u>be in</u> $\underline{\underline{V}}$. <u>Then</u> G <u>is a splitting group in</u> $\underline{\underline{V}}$ <u>if and only if it is a retract of some $\underline{\underline{V}}$-free group</u>.

<u>PROOF</u>: If G is a splitting group in $\underline{\underline{V}}$, then clearly every $\underline{\underline{V}}$-free presentation splits. Thus G is a retract of every $\underline{\underline{V}}$-free group of which it is a quotient. Conversely, let G be a retract of the $\underline{\underline{V}}$-free group F . If $f : K \twoheadrightarrow G$ is a surjective map in $\underline{\underline{V}}$, then we may define $g : F \to K$ such that the square

$$\begin{array}{ccc} F & \xrightarrow{\;h\;} & G \\ {\scriptstyle g}\,\downarrow & & \| \\ K & \xrightarrow{\;f\;} & G \end{array}$$

is commutative. If $s : G \to F$ is a right inverse of h then f splits by $gs : G \to K$.

It is clear that in a Schreier variety $\underline{\underline{V}}$ every retract of a $\underline{\underline{V}}$-free group is $\underline{\underline{V}}$-free, for it is isomorphic (under s) to a subgroup of a $\underline{\underline{V}}$-free group. In other varieties $\underline{\underline{V}}$ however there may exist splitting groups which are not $\underline{\underline{V}}$-free (see Theorem 4.6).

LEMMA 4.2. *Let* G *be a retract of the* $\underline{\underline{V}}$*-free group* F. *Then* G^q_{ab} *is a direct summand in* F^q_{ab}, *and* $V^qG = 0$ *for* $q = 0,1,2,\ldots$.

PROOF: Since $(-)^q_{ab}$ is a functor it follows that

$$G^q_{ab} \xrightarrow{s_*} F^q_{ab} \xrightarrow{h_*} G^q_{ab}$$

is the identity. Hence G^q_{ab} is a direct summand in F^q_{ab}. Also,

$$V^qG \xrightarrow{s_*} V^qF \xrightarrow{h_*} V^qG$$

is the identity. But $V^qF = 0$, hence $V^qG = 0$.

THEOREM 4.3. *Let* $\underline{\underline{V}}$ *be a variety of exponent* $q = 0$ *or* $q = p^n$, p *prime. Let* G *be a retract of a* $\underline{\underline{V}}$*-free group. Then there is a* $\underline{\underline{V}}$*-free group* F *and a homomorphism* $f : F \to G$ *such that* f *induces isomorphisms*

$$f_i : F/F^q_i \xrightarrow{\sim} G/G^q_i \ , \ q = 0,1,\ldots, \ i = 1,2,\ldots \ ,$$

and a monomorphism $f_\omega : F/F^q_\omega \rightarrowtail G/G^q_\omega$.

PROOF: This immediately follows from Corollary 1.3.

COROLLARY 4.4. *Let* $\underline{\underline{V}}$ *be a variety of exponent* $q = 0$ *or* $q = p^n$, p *prime, satisfying* (P_o). *Suppose that* G *is a retract of a* $\underline{\underline{V}}$*-free group such that* G_{ab} *is free in* $\underline{\underline{V}} \cap \underline{\underline{Ab}}$ *on a set of cardinality* α. *Then* G *contains a subgroup* F *which is* $\underline{\underline{V}}$*-free on a set of cardinality* α.

COROLLARY 4.5. *Let* $\underline{\underline{V}}$ *be a nilpotent variety of exponent* $q = 0$ *or* $q = p^n$, p *prime. Then a retract of a* $\underline{\underline{V}}$*-free group is* $\underline{\underline{V}}$*-free*.

THEOREM 4.6. *Let* $\underline{\underline{V}}$ *be a nilpotent variety of exponent* $q' = p_1^{n_1} p_2^{n_2} \ldots p_k^{n_k}$. *Then a retract* G *of a* $\underline{\underline{V}}$*-free group is of the form*

$$G = F_1 \times F_2 \times \ldots \times F_k$$

where F_i *is free in the variety* $\underline{\underline{V}}_i = \underline{\underline{V}} \cap B_{p_i^{n_i}}$, $i = 1,\ldots,k$.

PROOF: Since G is nilpotent and of exponent dividing q' it is a direct product of groups P_i of exponent dividing $p_i^{n_i}$, $i = 1,\ldots,k$. Since G is a retract of a $\underline{\underline{V}}$-free group F, $(P_i)_{ab}$ must be a direct summand in F_{ab}, so that $(P_i)_{ab}$ must be free in $\underline{\underline{V}}_i \cap \underline{\underline{Ab}}$. Let F_i be the free group in $\underline{\underline{V}}_i$ such that

$$(F_i)_{ab} \cong (P_i)_{ab} .$$

We may then define a map $f : F_1 \times F_2 \times \ldots \times F_k \to G$ satisfying the hypotheses of Corollary 1.2 for $q = 0$, thus proving our theorem.

IV.5. Parafree Groups

Let us say that two groups K,G have the same lower central sequence if there exist isomorphisms

(5.1) $$h_i : K/K_i^o \cong G/G_i^o \quad , \; i = 1,2,\ldots$$

such that for every $i \geq 2$ the square

$$\begin{array}{ccc} K/K_i^o & \twoheadrightarrow & K/K_{i-1}^o \\ h_i \downarrow & & h_{i-1} \downarrow \\ G/G_i^o & \twoheadrightarrow & G/G_{i-1}^o \end{array}$$

is commutative. Note that we do not require the existence of a homomorphism between K and G.

In [13], [14] G. Baumslag has defined a $\underline{\underline{V}}$-parafree group (of rank n) as a group G such that

(i) G is in $\underline{\underline{V}}$;

(ii) G is residually nilpotent;

(iii) G has the same lower central sequence as a $\underline{\underline{V}}$-free group (of rank n).

If $\underline{\underline{V}}$ is a variety satisfying (P_o) it is apparent that with respect to the lower central sequence $\underline{\underline{V}}$-parafree groups behave exactly like $\underline{\underline{V}}$-free groups. Baumslag has given many examples of $\underline{\underline{V}}$-parafree groups which are not $\underline{\underline{V}}$-free.

In this section we shall express part (iii) of Baumslag's definition of $\underline{\underline{V}}$-parafree groups in homological terms (Corollary 5.4) and then prove a number of propositions about parafree groups.

PROPOSITION 5.1. *Let* G *have the same lower central sequence as some* $\underline{V}$*-free group, and let* $g : G \to G/G_i^o$, $i \geq 2$ *denote the canonical projection. Then* $g_* : VG \to V(G/G_i^o)$ *is the zero map*.

PROOF: Given $i \geq 2$ there exists a $\underline{V}$-free group F and a homomorphism $f : F \to G$ inducing $h_{i+1} : F/F_{i+1}^o \xrightarrow{\sim} G/G_{i+1}^o$. Note that then f induces the isomorphisms

$$h_m : F/F_m^o \xrightarrow{\sim} G/G_m^o \quad , \quad 1 \leq m \leq i$$

The diagram

$$\begin{array}{ccccc} F_i^o/F_{i+1}^o & \rightarrowtail & F/F_{i+1}^o & \twoheadrightarrow & F/F_i^o \\ \downarrow k_i & & \downarrow h_{i+1} & & \downarrow h_i \\ G_i^o/G_{i+1}^o & \rightarrowtail & G/G_{i+1}^o & \twoheadrightarrow & G/G_i^o \end{array} \tag{5.2}$$

shows that k_i is an isomorphism. Consider the diagram

$$\begin{array}{ccccccccc} 0 & \to & V(F/F_i^o) & \to & F_i^o/F_{i+1}^o & \to & F_{ab} & \to & (F/F_i^o)_{ab} & \to 0 \\ \downarrow & & \downarrow (h_i)_* & & \downarrow k_i & & \downarrow h_2 & & \downarrow (h_i)_* & \\ VG & \xrightarrow{g_*} & V(G/G_i^o) & \to & G_i^o/G_{i+1}^o & \to & G_{ab} & \to & (G/G_i^o)_{ab} & \to 0 \end{array} \tag{5.3}$$

which is easily seen to be commutative. Since h_i is an isomorphism, $(h_i)_*$ also is. The fact that k_i is an isomorphism, then yields that $g_* : VG \to V(G/G_i^o)$ is the zero map.

The following Lemma 5.2 is an extension of our basic Theorem 1.1.

LEMMA 5.2. *Let* $f : K \to G$ *be a homomorphism of groups in* $\underline{V}$. *Suppose that* f *induces an isomorphism* $f_* : K_{ab}^q \xrightarrow{\sim} G_{ab}^q$. *Suppose also that* $g_* : V^qG \to V^q(G/G_i^q)$ *is the zero map for all* $i \geq 2$. *Then* f *induces an isomorphism*

$$f_i^q : K/K_i^q \xrightarrow{\sim} G/G_i^q \tag{5.4}$$

for every $i \geq 1$ and a monomorphism $f^q_\omega : H/H^q_\omega \rightarrowtail G/G^q_\omega$.

PROOF: The proof is the same as the proof of Theorem 1.1 , the only difference being that α_5 in diagram (1.6) is not necessarily an epimorphism. However, since in our case $g_* : V^qG \to V^q(G/G^q_i)$ is the zero map we may replace in diagram (1.6) the group V^qG by the trivial group. Then the argument goes through without any difficulty.

PROPOSITION 5.3. Let $\underline{\underline{V}}$ be a variety of exponent zero. Suppose that $g_* : VG \to V(G/G^0_i)$ is the zero map for $i \geq 2$. Then, if G_{ab} is free abelian there exists a $\underline{\underline{V}}$-free group F and a homomorphism $f : F \to G$ such that f induces isomorphisms

(5.5) $$f^q_i : F/F^q_i \xrightarrow{\sim} G/G^q_i \ , \ i = 1,2,\ldots \ ; \ q = 0,1,2,\ldots \ .$$

PROOF: First let $q = 0$. Choose elements $x_j \in G$, $j \in J$ such that their images in G_{ab} form a basis. Let F be the $\underline{\underline{V}}$-free group on (y_j) , $j \in J$ and define $f : F \to G$ by setting $f(y_j) = x_j$. By Lemma 5.2 we conclude that f induces isomorphisms $f^0_i : F/F^0_i \xrightarrow{\sim} G/G^0_i$, $i \geq 1$. The same argument works for $q \neq 0$ if we can show that

$$g^!_* : V^qG \to V^q(G/G^q_i)$$

is the zero map for all $i \geq 2$. But by definition $G^q_i \supseteq G^0_i$, so that $g^!_*$ factors as

(5.6) $$g^!_* : V^qG \xrightarrow{g^q_*} V^q(G/G^0_i) \to V^q(G/G^q_i) \ .$$

It is thus enough to show that $g^q_* = 0$. Since G_{ab} is free abelian we have by Corollary III.8.2

(5.7) $$V^qG = VG \otimes Z/qZ \ , \quad V^q(G/G^0_i) = V(G/G^0_i) \otimes Z/qZ$$

so that $g_*^q = g_* \otimes 1 = 0$.

Note that the part of the conclusion with $q = 0$ remains true if $\underline{\underline{V}}$ is an arbitrary variety and G_{ab} is free in $\underline{\underline{V}} \cap \underline{\underline{Ab}}$.

COROLLARY 5.4. Let G be in $\underline{\underline{V}}$. Then the following statements are equivalent.

(i) G has the same lower central sequence as some $\underline{\underline{V}}$-free group;

(ii) G_{ab} is free in $\underline{\underline{V}} \cap \underline{\underline{Ab}}$ and $g_* : VG \to V(G/G_i^0)$ is the zero map for all $i \geq 2$;

(iii) There exists a $\underline{\underline{V}}$-free group F and a homomorphism $f : F \to G$ such that $f_i^0 : F/F_i^0 \xrightarrow{\sim} G/G_i^0$, $i \geq 1$.

PROOF: This immediately follows from Propositions 5.1 and 5.3.

COROLLARY 5.5. Let G be a residually nilpotent group in $\underline{\underline{V}}$ with G_{ab} free in $\underline{\underline{V}} \cap \underline{\underline{Ab}}$ and $VG = 0$. Then G is $\underline{\underline{V}}$-parafree.

COROLLARY 5.6. Let $\underline{\underline{V}}$ be a variety of exponent zero, satisfying (P_0). If G is $\underline{\underline{V}}$-parafree (of rank n), then it contains a $\underline{\underline{V}}$-free group F (of rank n) such that the embedding $f : F \rightarrowtail G$ induces isomorphisms

(5.8) $$f_i^q : F/F_i^q \xrightarrow{\sim} G/G_i^q \ , \quad i = 1,2,\ldots \ , \quad q = 0,1,2,\ldots \ .$$

PROOF: From Proposition 5.3 and Lemma 5.2 it follows that there exists $f : F \to G$ with F free in $\underline{\underline{V}}$, inducing a monomorphism

(5.9) $$f_\omega^0 : F/F_\omega^0 \rightarrowtail G/G_\omega^0$$

Since $F_\omega^0 = e$, the map f_ω^0 factors as

(5.10) $$F \xrightarrow{f} G \to G/G_\omega^0$$

whence it follows that f itself is monomorphic. Obviously f has the required properties.

COROLLARY 5.7. Let $\underline{\underline{V}}$ be a variety of exponent zero, satisfying (P_p) for infinitely many primes. Let G be $\underline{\underline{V}}$-parafree and let (x_j), $j \in J$ be a set of elements in G, such that their images in G_{ab} are linearly independent. Then (x_j), $j \in J$ generates a $\underline{\underline{V}}$-free subgroup F of G.

PROOF: By Proposition I.3.1 it is enough to prove the assertion for all finite subsets (x_j), $j \in J'$. But for any finite set (x_j), $j \in J'$ we may find a prime p such that the images of (x_j), $j \in J'$ in G^p_{ab} are linearly independent and such that $\underline{\underline{V}}$ satisfies (P_p). We may then enlarge the set (x_j), $j \in J'$ to a set (x_j), $j \in J''$, $J' \subseteq J''$ such that the images in G^p_{ab} form a basis. We then prove that (x_j), $j \in J''$ freely generates a $\underline{\underline{V}}$-free subgroup by considering the $\underline{\underline{V}}$-free group F on (y_j), $j \in J''$ and the map $f : F \to G$ defined by $f(y_j) = x_j$. Lemma 5.2 then yields the desired result since $F^p_\omega = e$.

Note that the conclusion of Corollary 5.7 remains correct if the hypothesis that G be residually nilpotent is dropped from the statement. (It is implicitly part of the hypothesis that G be $\underline{\underline{V}}$-parafree.) Next we state two other immediate consequences of our Lemma 5.2.

PROPOSITION 5.8. Let $f : K \to G$ be a homomorphism in $\underline{\underline{V}}$. Suppose that G is $\underline{\underline{V}}$-parafree and that K is residually nilpotent. If $f^o_2 : K_{ab} \tilde{\to} G_{ab}$, then K is $\underline{\underline{V}}$-parafree (of the same rank as G) and f is a monomorphism.

COROLLARY 5.9. Let $f : K \to G$ be an epimorphism of $\underline{\underline{V}}$-parafree groups of the same finite rank. Then f is an isomorphism.

PROOF: We only have to prove that $f_* : K_{ab} \tilde{\to} G_{ab}$. But if f is surjective, so is f_*. Since K and G have the same finite rank, the groups K_{ab} and G_{ab} are free in $\underline{\underline{V}} \cap \underline{\underline{Ab}}$ of the same finite rank. Hence f_* is monomorphic.

We conclude with two results of a rather different nature.

PROPOSITION 5.10. *Let $\underline{\underline{V}}$ be a variety of exponent zero. A $\underline{\underline{V}}$-parafree group of rank $m \geq 1$ has a $\underline{\underline{V}}$-parafree quotient of rank $m-1$.*

PROOF: Choose $a \in G$ such that its image in G_{ab} is a basis element. Consider the normal subgroup N of G generated by a. The extension $N \overset{h}{\rightarrowtail} G \overset{f}{\twoheadrightarrow} Q$ yields the exact sequence

$$(5.11) \qquad VG \xrightarrow{f_*} VQ \xrightarrow{\delta_*} N/[G,N] \xrightarrow{h_*} G_{ab} \xrightarrow{f_*} Q_{ab} \to 0 .$$

Since N is generated by one element, $N/[G,N]$ is cyclic. Since rank $G_{ab} = m$ and rank $Q_{ab} = m-1$, the map h_* is non-trivial. Since G_{ab} is free abelian, h_* must be monomorphic. It follows that $f_* : VG \to VQ$ is epimorphic. Now consider the square $(i \geq 2)$

$$(5.12) \qquad \begin{array}{ccc} VG & \xrightarrow{f_*} & VQ \\ g_* \downarrow & & \downarrow g_*' \\ V(G/G_i^o) & \longrightarrow & V(G/G_i^o N) \end{array}$$

Since g_* is the zero map, the diagonal is the zero map. Since f_* is epimorphic, g_*' has to be the zero map, also. But of course $G/G_i^o N = Q/Q_i^o$. It then follows from Corollary 5.4 that Q has the same lower central sequence as a $\underline{\underline{V}}$-free group (of rank $m-1$). Clearly, Q/Q_ω^o is then $\underline{\underline{V}}$-parafree of rank $m-1$.

PROPOSITION 5.11. *Every $\underline{\underline{V}}$-parafree group of rank m is a quotient of a $\underline{\underline{V}}$-parafree group of rank $m+1$.*

PROOF: Let G be $\underline{\underline{V}}$-parafree of rank m. Consider $K = G *_V C$ where C is the $\underline{\underline{V}}$-free group on one generator. Clearly K_{ab} is free in

$\underline{\underline{V}} \cap \underline{\underline{Ab}}$ of rank $m+1$. Next we prove that $g^!_* : VK \to V(K/K^o_i)$ is the zero map for $i \geqslant 2$. Since $G/G^o_i \subseteq K/K^o_i$ we may consider the following diagram

$$
(5.13) \qquad
\begin{array}{ccc}
VG & \xrightarrow{\quad g_* \quad} & V(G/G_i) \\
\beta\uparrow & & \Big| \\
VG \oplus VC & & \Big| \\
\alpha\downarrow & & \downarrow \\
VK & \xrightarrow{\quad g^!_* \quad} & V(K/K_i)
\end{array}
$$

where α is an isomorphism by Theorem III.6.1. Since $VC = 0$, the map β is an isomorphism. Thus, the fact that g_* is the zero map, implies that $g^!_*$ is the zero map. It then follows from Corollary 5.4 that K has the same lower central sequence as a $\underline{\underline{V}}$-free group of rank $m+1$. Clearly K/K^o_ω is then $\underline{\underline{V}}$-parafree of rank $m+1$, and has G as quotient.

IV.6. The Deficiency

In this section we suppose that $\underline{\underline{V}}$ is a variety of exponent zero, and all groups considered will be finitely presentable in $\underline{\underline{V}}$.

Let the group G in $\underline{\underline{V}}$ be given by a finite $\underline{\underline{V}}$-presentation

$$
(6.1) \qquad G = gp_V(x_1,x_2,\ldots,x_n \mid y_1,y_2,\ldots,y_r) ,
$$

i.e. G is the quotient of the $\underline{\underline{V}}$-free group F on $x_1,x_2,\ldots,x_n$ by the normal subgroup R generated by $y_1,y_2,\ldots,y_r$. The elements $x_1,x_2,\ldots,x_n$ are called <u>generators</u>, and the elements $y_1,y_2,\ldots,y_r$ <u>relators</u>. The number $n-r$ is called the <u>deficiency</u> of the presentation. We define $def_V G$, the $\underline{\underline{V}}$-<u>deficiency</u> of the group G to be the maximum deficiency of the finite $\underline{\underline{V}}$-presentations of G . It will be a consequence of Theorem 6.1 that the maximum of the deficiencies of the finite $\underline{\underline{V}}$-presentations indeed exists and is finite. In particular there always exists a $\underline{\underline{V}}$-presentation whose deficiency is just $\mathrm{def}_V G$. If M is any finitely generated abelian group, we shall denote the minimum number of generators of M by sM . The following is a generalization of a theorem of P. Hall (see Epstein [27]) and a theorem of Knopfmacher [51].

THEOREM 6.1.

(6.2) (i) $\quad \mathrm{def}_V G \leq \mathrm{rank}\ G_{ab} - sVG$,

(6.3) (ii) $\quad \mathrm{def}_V G \leq \dim G_{ab}^p - \dim V^p G$, $\quad p$ prime.

PROOF: We only prove the first inequality, the proof of the second being similar. Let $R \overset{h}{\rightarrowtail} F \overset{g}{\twoheadrightarrow} G$ be a $\underline{\underline{V}}$-presentation of G with n generators and r relators. Consider the exact sequence

$$(6.4) \qquad 0 \to VG \to R/[F,R] \xrightarrow{h_*} F_{ab} \to G_{ab} \to 0 .$$

The image K of h_* is a free abelian group, since F_{ab} is free abelian. Thus $R/[F,R] \cong VG \oplus K$. Since $R/[F,R]$ is generated by the canonical images of the relators we get the inequality

$$(6.5) \qquad \begin{aligned} r \geq s(R/[F,R]) &= sVG + \mathrm{rank}\ K \\ &= sVG + \mathrm{rank}\ F_{ab} - \mathrm{rank}\ G_{ab} \\ &= sVG + n - \mathrm{rank}\ G_{ab} , \end{aligned}$$

thus proving statement (i).

We claim that for $\underline{\underline{V}} = \underline{\underline{Ab}}$, the $\underline{\underline{V}}$-deficiency is nothing else but the (torsion free) rank. To prove this let G be a finitely generated abelian group and let $R \rightarrowtail F \twoheadrightarrow G$ be a finite presentation in $\underline{\underline{Ab}}$. Tensoring with Q and counting the dimensions of the resulting vectorspaces yields

$$n-r = \text{rank } G .$$

It follows that in this case (6.2) is always an equality. We claim that (6.3) also is an equality. By Corollary III.8.2, we have for p a prime

$$\begin{aligned} \dim G^p_{ab} - \dim V^p G &= \\ &= (\dim G \otimes Z/pZ - \dim \mathrm{Tor}_1^Z(G,Z/pZ)) - \dim(VG \otimes Z/pZ) \\ &= \text{rank } G_{ab} - 0 \\ &= \mathrm{def}_V G , \end{aligned} \tag{6.6}$$

since $VG = \mathrm{Tor}_1^Z(Z,G)$ by (III.2.14).

<u>DEFINITION</u>: A (finitely presentable) group G in $\underline{\underline{V}}$ for which (6.2) is equality is called <u>efficient</u> or 0-<u>efficient</u> in $\underline{\underline{V}}$. A (finitely presentable) group G in $\underline{\underline{V}}$ for which (6.3) is an equality is called p-<u>efficient</u> in $\underline{\underline{V}}$.

<u>PROPOSITION 6.2</u>. <u>Let</u> G <u>be a finitely presentable group in</u> $\underline{\underline{V}}$. <u>Then we have for all primes</u> p

$$\text{rank } G_{ab} - sVG \leq \dim G^p_{ab} - \dim V^p G , \tag{6.7}$$

<u>and there always exists a prime</u> p <u>for which we have equality</u>.

<u>PROOF</u>: Since $G^p_{ab} = G_{ab} \otimes Z/pZ$ and

$$V^p G \cong VG \otimes Z/pZ \oplus \mathrm{Tor}_1^Z(G_{ab}, Z/pZ) ,$$

(see Corollary III.8.2), we obtain

$$\begin{aligned} \dim G^p_{ab} - \dim V^p G &= (\dim(G_{ab} \otimes Z/pZ) - \dim \mathrm{Tor}_1^Z(G_{ab}, Z/pZ)) - \dim(VG \otimes Z/pZ) \\ &= \text{rank } G_{ab} - \dim(VG \otimes Z/pZ) \\ &\geq \text{rank } G_{ab} - sVG . \end{aligned} \tag{6.8}$$

Moreover, VG may be written as

$$VG = Z\oplus\ldots\oplus Z\oplus Z/n_1Z\oplus Z/n_2Z\oplus\ldots\oplus Z/n_kZ$$

with n_i/n_{i+1} for $1 \leq i < k$. Consequently we have equality if $k = 0$ or if $k \neq 0$ and p/n_1 .

Next we give some examples.

(i) We have already seen (see (6.6)) that in $\underline{\underline{V}} = \underline{\underline{Ab}}$ every finitely generated group is efficient in $\underline{\underline{V}}$ and also p-efficient in $\underline{\underline{V}}$ for every prime p .

(ii) Consider $\underline{\underline{V}} = \underline{\underline{Gr}}$.

PROPOSITION 6.3. Every finitely generated abelian group G is efficient in $\underline{\underline{V}} = \underline{\underline{Gr}}$.

PROOF: Let $G \cong Z \oplus Z \oplus \ldots \oplus Z \otimes Z_{n_1} \oplus \ldots \oplus Z_{n_k}$ with n_i/n_{i+1} , $1 \leq i < k$. Choose for every direct summand a generator x_ℓ , $1 \leq \ell \leq \text{rank } G + k = m$. We need $\binom{m}{2}$ relators to make the x_ℓ commute and k relators to make the x_ℓ of finite order, rank $G+1 \leq \ell \leq m$. On the other hand we have

$$(6.9) \qquad \begin{aligned} \text{rank } G_{ab} &= \text{rank } G , \\ sH_2G &= \binom{m}{2} . \end{aligned}$$

The second equation being easily deduced from the Künneth Theorem (see II.5.13). It follows that we indeed get equality in (6.2).

(iii) Let $\underline{\underline{V}}$ be any variety of exponent zero.

PROPOSITION 6.4. Every one relator group

$$(6.10) \qquad G = gp_V(x_1, x_2, \ldots, x_n | y)$$

is efficient in $\underline{\underline{V}}$.

PROOF: Let $R \overset{h}{\rightarrowtail} F \overset{g}{\twoheadrightarrow} G$ be the given $\underline{\underline{V}}$-presentation. The associated 5-term sequence reads

$$0 \to VG \xrightarrow{\delta_*} R/[F,R] \xrightarrow{h_*} F_{ab} \xrightarrow{g_*} G_{ab} \to 0 .$$

Since $R/[F,R]$ is generated by the canonical image of y, it is cyclic. We have to consider the two cases: $y \in F_2^0$ and $y \notin F_2^0$. If $y \notin F_2^0$ then h_* is non trivial. Since F_{ab} is free abelian, h_* must be monomorphic. Hence $VG = 0$. If $y \in F_2^0$, then $g_* : F_{ab} \xrightarrow{\sim} G_{ab}$ and $\delta_* : VG \xrightarrow{\sim} R/[F,R]$. In both cases we have

(6.11) $$\text{rank } G_{ab} - sVG = n-1$$

so that G is efficient in $\underline{\underline{V}}$.

(iv) Let $\underline{\underline{V}}$ be any variety of exponent zero. We will show in Section 7 that any group G in $\underline{\underline{V}}$ given by a presentation with $n+r$ generators and r relators where $sG_{ab} \leq n$ is efficient in $\underline{\underline{V}}$.

(v) It is easy to see that any free nilpotent group is efficient in $\underline{\underline{V}} = \underline{\underline{Gr}}$.

(vi) In [17] Beyl has shown that finite groups that are extensions of a cyclic group by a cyclic group are efficient in $\underline{\underline{Gr}}$.

(vii) In [81] Swan has shown that the following groups G are <u>not</u> efficient in $\underline{\underline{V}} = \underline{\underline{Gr}}$. Consider the elementary abelian group A of order 7^k and the cyclic group C_3 of order 3 with generator x. Define the action of C_3 on A by

$$xa = a^2 \quad , \quad a \in A .$$

Let $G = A \rtimes C_3$. It may be shown that $\text{def}_V G$ tends to $-\infty$ as k tends to $+\infty$, and that $H_2^0 G = 0$.

It seems to be an open question whether nilpotent groups are efficient in $\underline{\underline{Gr}}$. However we have

THEOREM 6.5. *Let* G *be a group in* $\underline{\underline{V}}$ *, given by a finite* $\underline{\underline{V}}$*-presentation. Then there exists a p-efficient group* K *in* $\underline{\underline{V}}$ *and a surjective homomorphism* $f : K \to G$ *which induces an isomorphism*

(6.12) $$f_i : K/K_i^p \tilde{\to} G/G_i^p$$

for every $i \geq 1$.

PROOF: We give the proof for $p = 0$, the proof for p a prime being similar. Consider the finite $\underline{\underline{V}}$-presentation $R \overset{h}{\rightarrowtail} F \overset{g}{\twoheadrightarrow} G$ and the corresponding 5-term sequence

$$0 \to VG \xrightarrow{\delta_*} R/[F,R] \xrightarrow{h_*} F_{ab} \xrightarrow{g_*} G_{ab} \to 0 .$$

Denote by I the image of h_* . Suppose the generators of the presentation are $x_1, x_2, \ldots, x_n$. Then we define the group K by

(6.13) $$K = gp_V(x_1, x_2, \ldots, x_n | y_1, \ldots, y_j, z_1, \ldots, z_k)$$

where

(i) $y_1, y_2, \ldots, y_j, z_1, z_2, \ldots, z_k$ are elements in R ,

(ii) the canonical images of $y_1, \ldots, y_j$ in I form a basis of I ,

(iii) the images of $z_1, \ldots, z_k$ in $R/[F,R]$ form a set of generators of $\delta_*(VG)$ with $k = sVG$.

Obviously there is a surjective map $f : K \to G$. Also, it is clear that f induces an isomorphism $f_* : K_{ab} \tilde{\to} G_{ab}$ and an epimorphism $f_* : VK \to VG$. The assertion about the lower central series then follows from Theorem 1.1. It remains to show that K is efficient in $\underline{\underline{V}}$. We have

(6.14) $j+k = sVG + n-\text{rank } G_{ab}$

and therefore

(6.15) $\text{rank } G_{ab}-sVG = n-(j+k) \leq \text{def}_V K \leq$

$\leq \text{rank } K_{ab}-sVK \leq \text{rank } G_{ab}-sVG$

where the last inequality follows from the fact that $f_* : VK \to VG$ is epimorphic. It follows that K is efficient.

COROLLARY 6.6. *In a nilpotent variety of exponent zero every (finitely generated) group is efficient.*

PROOF: This is clear from Theorem 6.5 and (6.12).

We note that by Corollary 6.6 we have, for a nilpotent variety $\underline{\underline{V}}$ of exponent zero, a homological characterization of the $\underline{\underline{V}}$-deficiency, namely

(6.16) $\text{def}_V G = \text{rank } G_{ab}-sVG$.

We conclude with the following result of Chen [18].

COROLLARY 6.7. *Let* G *be a group given by a* $\underline{\underline{Gr}}$*-presentation with* $n+r$ *generators and* $r+k$ *relators, where* $n = sG_{ab}$. *Given* $d \geq 1$ *there exists a group* K *with* n *generators and* k *relators with* $K/K_d^o \cong G/G_d^o$.

PROOF: The emphasis is on the fact that k relators suffice for a presentation of K. The deficiency of G/G_d in the variety $\underline{\underline{N}}_{d+1}$ of all nilpotent groups of class $\leq d-1$ is at least the deficiency of G in $\underline{\underline{Gr}}$, which turns out to be

$$\text{def}_{Gr} G \leq n+r-(r+k) = n-k .$$

Since G/G_d is generated by $sG_{ab} = n$ elements we may apply Corollary 6.6 to obtain an $\underline{\underline{N}}_{d-1}$-presentation of G/G_d by n generators and k relators. Taking as relators inverse images of these k relators in the absolutely free group on the same generators we obtain a $\underline{\underline{Gr}}$-presentation of K .

IV.7. Groups Given by Special Presentations

In this section we consider a variety $\underline{\underline{V}}$ satisfying (P_p) for $p = 0$ and all primes p . Note that $\underline{\underline{V}}$ is necessarily of exponent zero.

THEOREM 7.1. *Let G be a group with a $\underline{\underline{V}}$-presentation by $n+r$ generators and r relators. Suppose that G_{ab} is generated by n elements. Then*

(i) G_{ab} *is free abelian of rank* n .

(ii) $VG = 0$.

(iii) $\mathrm{def}_V G = n$; *in particular* G *is efficient in* $\underline{\underline{V}}$.

(iv) *Let p be a prime or $p = 0$. If $x_1,\ldots,x_n$ are elements of G , whose images in G^p_{ab} form a basis, then they generate a $\underline{\underline{V}}$-free subgroup F of G whose embedding $f : F \rightarrowtail G$ induces an isomorphism*

$$f^p_i : F/F^p_i \cong G/G^p_i \quad , \quad i = 1,2,\ldots .$$

PROOF: By Theorem 6.1 we have

$$n+r-r = n \leqslant \mathrm{def}_V G \leqslant \mathrm{rank}\ G_{ab} - sVG \leqslant n - sVG$$

so that we may conclude

$def_V G = n$;

$sVG = 0$, i.e. $VG = 0$;

rank $G_{ab} = n$.

The remaining assertion then follows from Theorem 2.2 since $V^p G = 0$ for all primes p .

COROLLARY 7.2. *Let* G *be a group with* $n+r$ *generators and* r *relators. Suppose that* G_{ab} *is generated by* n *elements. If* $x_1, x_2, \ldots, x_m$ *are elements in* G *whose images in* G_{ab} *are linearly independent, then they generate a* $\underline{\underline{V}}$*-free subgroup* F *of* G .

PROOF: By Theorem 7.1 we know that G_{ab} is free abelian of rank n . We may thus find a prime p such that the images of $x_1, \ldots, x_m$ in G^p_{ab} are linearly independent, hence are part of a basis. The result then easily follows from Theorem 7.1 (iv).

COROLLARY 7.3. (Magnus [60]) *Let* G *be a group with* $n+r$ *generators and* r *relators. Suppose* G *can be generated by* n *elements. Then* G *is* $\underline{\underline{V}}$*-free on* n *generators*.

PROOF: The group G_{ab} is generated by n elements, hence it is free abelian of rank n . If $x_1, x_2, \ldots, x_n$ generate G ; their images generate G_{ab} , hence form a basis of G_{ab} . The result then follows from Theorem 7.1 since $F = G$.

COROLLARY 7.4. *Let* $\underline{\underline{V}} = \underline{\underline{Gr}}$ *and let* G *be a knot group. Then for any abelian group* A , $H_2(G,A) = 0$.

PROOF: It is well-known that $G_{ab} = Z$ (see [19]). It is thus enough to prove that $H_2G = 0$. But G has a presentation by $1+r$ generators and r relators (see [19]), so that by Theorem 7.1 we have $VG=H_2G=0$.

THEOREM 7.5. *Let G be a group with a $\underline{\underline{V}}$-presentation by $n+r$ generators and r relators. Suppose that G^p_{ab}, p prime is generated by n elements. Then*

(i) $\dim G^p_{ab} = n$.

(ii) $V^p G = 0$.

(iii) $\mathrm{def}_V G = n$; in particular G is efficient in $\underline{\underline{V}}$.

(iv) *If $x_1, x_2, \ldots, x_n$ are elements of G, whose images in G^p_{ab} form a basis, then they generate a $\underline{\underline{V}}$-free subgroup F of G, whose embedding $f : F \rightarrowtail G$ induces an isomorphism*

$$f^p_i : F/F^p_i \xrightarrow{\sim} G/G^p_i$$

for every $i \geqslant 1$.

PROOF: The proof is the same as the proof of Theorem 7.1.

COROLLARY 7.6. *Let $G = gp_V(a_1, a_2, \ldots, a_n, b_1, b_2, \ldots, b_r \mid y_1, \ldots, y_r)$ where $y_k = b_k \cdot \varrho_k(a_1, a_2, \ldots, a_n, b_1, b_2, \ldots, b_r)$, $k = 1, \ldots, r$ and ϱ_k denotes words in $a_1, a_2, \ldots, a_n, b_1, b_2, \ldots, b_r$. Suppose that ϱ_k has exponent sum $p \cdot t_k$ in b_k ($p = 0$ or p a prime). Then $a_1, \ldots, a_n$ generate a $\underline{\underline{V}}$-free subgroup F of rank n and the embedding $f : F \rightarrowtail G$ induces an isomorphism*

$$f^p_i : F/F^p_i \xrightarrow{\sim} G/G^p_i$$

for every $i \geqslant 1$.

PROOF: It is apparent that G^p_{ab} is generated by the images of $a_1, \ldots, a_n$. Thus G and the elements $a_1, \ldots, a_n$ in G satisfy the hypotheses of Theorem 7.1 ($p = 0$) or Theorem 7.5 (p a prime).

COROLLARY 7.7. Let $G = gp_V(a_1,\dots,a_n,b|y)$ where the exponent sum s of b in y is non-zero. Then $a_1,\dots,a_n$ generate a free subgroup F of G. If p is a prime with $p \nmid s$, then the embedding $f : F \rightarrowtail G$ induces an isomorphism

$$f_i^p : F/F_i^p \xrightarrow{\sim} G/G_i^p$$

for every $i \geq 1$.

PROOF: If p is a prime with $p \nmid s$, then G_{ab}^p is generated by the images of $a_1,\dots,a_n$. Theorem 7.5 then yields the result.

Note that for $\underline{\underline{V}} = \underline{\underline{Gr}}$ the first part of the assertion follows from the so called "Freiheitssatz" of Magnus [58].

IV.8. The Huppert-Thompson-Tate Theorem

In this section we consider the variety $\underline{\underline{V}} = \underline{\underline{Gr}}$.

LEMMA 8.1. Let $f : K \to G$ be the embedding of a subgroup $K \subseteq G$ of finite index n, say. Let $q > 0$ be a number with $(q,n) = 1$. Then $f_* : H_i^q K \to H_i^q G$ is epimorphic.

PROOF: Since $[G:K] < \infty$, the corestriction map exists. Thus by [43], Theorem VI.16.4 we have that for $i \geq 1$

$$f_* \circ \mathrm{Cor} : H_i^q G \to H_i^q K \to H_i^q G$$

is just multiplication by n. But since $(q,n) = 1$ this is an isomorphis. It follows that f_* must be epimorphic.

THEOREM 8.2. Let K be a subgroup of G of index n, say. Suppose $p \nmid n$, p prime. If for $q = p^k$ we have $f_* : K_{ab}^q \xrightarrow{\sim} G_{ab}^q$, then f induces an isomorphism

$$f_i^q : K/K_i^q \tilde{\to} G/G_i^q$$

<u>for every</u> $i \geq 1$.

<u>PROOF</u>: This is an easy consequence of Theorem 1.1 since by Lemma 8.1 the map $f_* : H_2^q K \to H_2^q G$ is epimorphic.

We now easily get the Huppert-Thompson-Tate Theorem on normal p-complements.

<u>COROLLARY 8.3</u>. <u>Suppose</u> G <u>finite. Let</u> K <u>be a Sylow-p-subgroup of</u> G. <u>Suppose that</u> $f : K_{ab}^p \tilde{\to} G_{ab}^p$. <u>Then there is a normal subgroup</u> N <u>of</u> G <u>such that</u> $K \cap N = e$ <u>and</u> $G/N \cong K$ (N <u>is called a normal</u> p-<u>complement</u>).

<u>PROOF</u>: (Stallings [74]) Apply Theorem 8.2 with $q = p$. Since K is a p-group there exists n with $K_n^p = e$. Set $N = G_n^p$.

CHAPTER V

CENTRAL EXTENSIONS

In this chapter we have assembled some results on central extensions. We have tried to focus our attention on those results of group theoretical interest which are obtained from the 5-term sequence or the Ganea term (Section 2). In particular we have refrained from stating results that require spectral sequence techniques for their proofs.

Let $E : N \rightarrowtail G \twoheadrightarrow Q$ be a central extension. In Section 1 we exhibit a module structure in H_*G over the Pontrjagin ring of N. In Section 2 we use this module structure to yield the so-called Ganea term. Also, an explicit description of the Ganea term by means of free presentations is given. The Ganea term enables us to extend the 5-term sequence in integral homology associated with E by one term to the left. In Section 3 we define various classes of central extensions using properties of $\Pi\Delta[E] : H_2Q \to N$ associated with E via the universal coefficient sequence. Section 4 relates properties of the extension E with properties of the module structure in H_*G. In Section 5 we take a closer look at stem extensions and stem covers as defined in Section 3, and discuss in Section 6 the special case of perfect groups. The results presented in these two sections constitute in principle the theory of Schur. In Section 7 that theory is generalized by replacing the integral homology by homology with coefficients in Z/qZ. The theory of Schur and its generalization then yield in Section 8 a homological characterization of terminal and unicentral groups. In the last two sections we present applications of the Ganea term as obtained in Section 2. First we obtain some of the better known theorems on the order of the Schur multiplicator of a finite nilpotent group. Also, an estimate on the rank of H_2G for G nilpotent is obtained. Finally in Section 10 we prove some theorems of Hall's type, i.e. theorems that yield information about the group K if we are given information about $K/[H,H]$ for H a nilpotent normal subgroup of K.

The content of this chapter stems from a large number of papers and people. First of all we have to mention the famous papers [72], [73] where Schur has introduced and studied H_2G with his bare hands, and proved many of the results of Section 3, 5, 6 and more. The presentation we have adopted here is on the lines of Eckmann-Hilton-Stammbach [22]. We have also used ideas from P. Hall [37], Gruenberg [35] and G. Rinehart [69]. The Ganea term (Section 2) was first obtained by Ganea [30] by topological methods, its description by means of free presentation is from Eckmann-Hilton-Stammbach [22]. The module structure in H_*G (Sections 1, 4) does not seem to appear in the literature. The generalization of the theory of Schur and its applications (Sections 7, 8) are due to Evens [28], where other applications are to be found, also. Our presentation is simpler than Evens' since we avoid the use of spectral sequences. The estimates on the order of H_2G (Section 9) for G finite nilpotent are due to Green [33] and Gaschütz - Neubüser - Yen [31]; the proof of Proposition 9.5 is due to Vermani [84]. The content of Section 10 is to be found in Robinson [70], where further applications are given, also. The relation of the approach presented in [70] with homology theory, in particular with the Ganea term, seems to have escaped Robinson.

Finally we would like to mention some papers in which results presented here are extended. In Johnson [48] and Leedham-Green [53] parts of the theory of Schur are generalized to varieties. In Eckmann-Hilton-Stammbach [24] the Ganea term is used to yield information about the Schur multiplicator of a central quotient of a direct product of groups. In

[21], [23], [36], [44], [67], [80] the sequence (2.7) is extended to the left and also generalized to non-central extensions.

The following papers are related to the content of the indicated sections: Kervaire [49], Sections 3, 5, 6; Iwahori-Matsumoto [47], Section 2; Vermani [85], Section 9.

V.1. Generalities

In this section we shall prove some general results on the structure of the integral homology of a central extension. Consider a central extension

(1.1) $$E : N \overset{h}{\rightarrowtail} G \overset{g}{\twoheadrightarrow} Q ,$$

i.e. an extension with $N \subseteq ZG$. We may define a map $m : N\times G \to G$ by

(1.2) $$m(u,x) = u\cdot x \quad , \quad u \in N , \; x \in G .$$

Since N is central, m is a homomorphism. It is clear that m makes the following diagram commutative

(1.3) $$\begin{array}{ccccc} N & \rightarrowtail & N\times G & \twoheadrightarrow & G \\ \| & & \downarrow m & & \downarrow g \\ N & \overset{h}{\rightarrowtail} & G & \overset{g}{\twoheadrightarrow} & Q \end{array}$$

The map m induces $m_* : H_*(N\times G) \to H_*G$ and by the Künneth Theorem (see II.5.13) a homomorphism

(1.4) $$\mu : H_*N \otimes H_*G \rightarrowtail H_*(N\times G) \xrightarrow{m_*} H_*G .$$

Here we use the notation H_*G to denote the graded group $\{H_iG\}$. Note that μ is a homomorphism of graded groups of degree zero. Of course, μ is natural; more precisely, if

$$
(1.5)\qquad \begin{array}{ccccc} N & \rightarrowtail & G & \twoheadrightarrow & Q \\ f_1\downarrow & & f_2\downarrow & & f_3\downarrow \\ N' & \rightarrowtail & G' & \twoheadrightarrow & Q' \end{array}
$$

is a map of central extensions, then the diagram

$$
(1.6)\qquad \begin{array}{ccc} H_*N \otimes H_*G & \xrightarrow{\mu} & H_*G \\ f_{1*}\otimes f_{3*}\downarrow & & \downarrow f_{2*} \\ H_*N'\otimes H_*G' & \xrightarrow{\mu'} & H_*G' \end{array}
$$

is commutative.

In Proposition 1.1 we shall give an explicit description of μ in terms of the inhomogeneous standard resolution. Let us first briefly recall the definition of that resolution; for more details see [43], pp.216-217. The (non-normalized) inhomogeneous standard resolution $\underline{\bar{B}}'(G)$ is a ZG-free resolution of Z

$$
\underline{\bar{B}}'(G) \ : \ \dots \to \bar{B}'_n \xrightarrow{\partial_n} \bar{B}'_{n-1} \xrightarrow{\partial_{n-1}} \dots \xrightarrow{\partial_1} \bar{B}'_o \xrightarrow{\partial_o} Z \to 0 \ .
$$

$\bar{B}'_n$ is the free ZG-module on all $[x_1|x_2|\dots|x_n]$, $x_i \in G$ and the differential ∂_n , $n \geq 1$ is defined by

$$
\begin{aligned}
\partial_n[x_1|\dots|x_n] &= x_1[x_2|\dots|x_n] + \\
&+ \sum_{i=1}^{n-1} (-1)^i [x_1|\dots|x_i x_{i+1}|\dots|x_n] + \\
&+ (-1)^n [x_1|\dots|x_{n-1}] \ .
\end{aligned}
$$

Note that $\bar{B}'_o \cong ZG$; the map ∂_o is then defined to be the augmentation $\varepsilon : ZG \to Z$.

Also we shall need to define <u>shuffles</u>. An (i,j)-shuffle π of the set $S = (1,\dots,i,i+1,\dots,i+j)$ is a permutation π of S with $\pi(k) < \pi(\ell)$

whenever $k < \ell \leq i$ or $i < k < \ell$. By Ω we denote the set of all (i,j)-shuffles. The signature $s(\pi)$ of an (i,j)-shuffle π is the signature of the permutation π . With this notation we may describe the map μ as follows.

PROPOSITION 1.1. *The map* μ *is induced by the chain transformation*

$$\mu([x_1|x_2|\ldots|x_i] \otimes [x_{i+1}|x_{i+2}|\ldots|x_{i+j}]) = \tag{1.7}$$
$$= \sum_{\pi\in\Omega} (-1)^{s(\pi)} [x_{\pi(1)}|x_{\pi(2)}|\ldots|x_{\pi(i+j)}]$$

where $x_1,\ldots,x_i \in N$ *and* $x_{i+1},\ldots,x_{i+j} \in G$.

PROOF: $H_*G = H_*(Z \otimes_G \underline{\bar{B}}'(G))$. We shall use, for short, the notation $\underline{C}G = Z \otimes_G \underline{\bar{B}}'(G)$. The map μ is then induced by the following composition of chain maps

$$\underline{Z}N\otimes\underline{Z}G \xrightarrow{\zeta} \underline{C}N\otimes\underline{C}G \xrightarrow{\gamma} \underline{C}N\times\underline{C}G \xrightarrow{\alpha} \underline{C}(N\times G) \xrightarrow{m_*} \underline{C}G \tag{1.8}$$

where $\underline{C}N\times\underline{C}G$ denotes the cartesian product of $\underline{C}N$ and $\underline{C}G$ (see [57], p.288 ff.), which is isomorphic to $\underline{C}(N\times G)$. Also, we have denoted by $\underline{Z}(-)$ the subcomplex of cycles in $\underline{C}(-)$; the map ζ is induced by the embeddings $\underline{Z}N \hookrightarrow \underline{C}N$ and $\underline{Z}G \hookrightarrow \underline{C}G$. We recall that ζ induces the embedding

$$\zeta_* : H_*(\underline{C}N) \otimes H_*(\underline{C}G) \rightarrowtail H_*(\underline{C}N \otimes \underline{C}G) \tag{1.9}$$

in the Künneth sequence (see [43], p.172 ff.).

Finally γ is the Eilenberg-Zilber map, i.e. the homotopy inverse of the Alexander-Whitney map; it is given as follows (see [57], p.243/248). Let $x_1,\ldots,x_i \in N$ and $x_{i+1},\ldots,x_{i+j} \in G$. Then

$$\gamma([x_1|\dots|x_i] \otimes [x_{i+1}|\dots|x_{i+j}]) =$$

(1.10)

$$\sum_{\pi\in\Omega} (-1)^{s(\pi)} [x_{\tau(1)}|\dots|x_{\tau(i+j)}] \otimes [x_{\sigma(1)}|\dots|x_{\sigma(i+j)}]$$

where, for $1 \leq k \leq i+j$ and $\pi \in \Omega$

$$x_{\tau(k)} = \begin{cases} x_{\pi(k)} & \text{, for } 1 \leq \pi(k) \leq i \text{ ,} \\ e & \text{, otherwise} \end{cases}$$

$$x_{\sigma(k)} = \begin{cases} x_{\pi(k)} & \text{, for } i+1 \leq \pi(k) \leq i+j \text{ ,} \\ e & \text{, otherwise} \end{cases}$$

It is then clear that

$$m\alpha\gamma([x_1|\dots|x_i] \otimes [x_{i+1}|\dots|x_{i+j}]) =$$

(1.11) $$= m(\sum_{\pi\in\Omega} (-1)^{s(\pi)} [(x_{\tau(1)},x_{\sigma(1)})|(x_{\tau(2)},x_{\sigma(2)})|\dots|(x_{\tau(i+j)},x_{\sigma(i+j)}])$$

$$= \sum_{\pi\in\Omega} (-1)^{s(\pi)} [x_{\pi(1)}|\dots|x_{\pi(i+j)}]$$

since, for each k, either $x_{\tau(k)}$ or $x_{\sigma(k)}$ is e.

If we consider, for N abelian, the extension

(1.12) $$N \rightarrowtail N \twoheadrightarrow e \text{ ,}$$

then (1.4) yields a map

(1.13) $$\mu_N : H_*N \otimes H_*N \to H_*N \text{ .}$$

It follows from the properties of the Künneth Theorem and the map γ (1.10) of the Eilenberg-Zilber Theorem (see [57], p.242) that the map μ in (1.13) is associative and commutative (in the graded sense). Moreover, the diagram

$$(1.14)\qquad \begin{array}{ccccc} e & \rightarrowtail & N & \twoheadrightarrow & N \\ \downarrow & & \| & & \downarrow \\ N & \rightarrowtail & N & \twoheadrightarrow & e \end{array}$$

yields, by naturality (see (1.6)), the diagram

$$(1.15)\qquad \begin{array}{ccc} H_*(e) \otimes H_*N & \xrightarrow{\mu=1} & H_*N \\ \downarrow & & \| \\ H_*N \otimes H_*N & \xrightarrow{\mu_N} & H_*N \end{array}$$

Since $H_*(e)$ is concentrated in dimension zero, and $H_0(e) = Z$, it follows that the embedding $H_*(e) \to H_*N$ is a unit for the map μ . We have thus proved

PROPOSITION 1.2. *For any abelian group* N , *the map* μ *makes* H_*N *into an associative and commutative graded ring with unit.*

We remark that, obviously, the ring H_*N can be identified with the Pontrjagin ring of the Eilenberg-MacLane space $K(N,1)$. For the central extension (1.1) we have

PROPOSITION 1.3. *The map* $\mu : H_*N \otimes H_*G \to H_*G$ *makes* H_*G *into a graded module over the ring* H_*N .

PROOF: Naturality applied to the diagram

$$(1.16)\qquad \begin{array}{ccccc} e & \rightarrowtail & G & \twoheadrightarrow & G \\ \downarrow & & \| & & \downarrow g \\ N & \rightarrowtail & G & \xrightarrow{g} & Q \end{array}$$

yields

(1.17)
$$\begin{array}{ccc} H_*(e) \otimes H_*G & \xrightarrow{\mu=1} & H_*G \\ \downarrow & & \| \\ H_*N \otimes H_*G & \xrightarrow{\mu} & H_*G \end{array}$$

so that μ is unitary. Associativity, i.e. the fact that the diagram

(1.18)
$$\begin{array}{ccc} H_*N \otimes H_*N \otimes H_*G & \xrightarrow{\mu_N \otimes 1} & H_*N \otimes H_*G \\ {\scriptstyle 1\otimes\mu}\downarrow & & \downarrow{\scriptstyle \mu} \\ H_*N \otimes H_*G & \xrightarrow{\mu} & H_*G \end{array}$$

is commutative, again follows from associativity of the Künneth Theorem and of the map γ (1.10) of the Eilenberg-Zilber Theorem (see [57], p.242).

PROPOSITION 1.4. _Let_ B, C _be torsion-free abelian groups. Then_ $H_*(B\times C) \cong H_*B \otimes H_*C$ _as rings_.

PROOF: We first remark that the integral homology of a torsion-free abelian group is clearly torsion-free. The Künneth Theorem then yields an isomorphism

(1.19)
$$H_*B \otimes H_*C \xrightarrow{\cong} H_*(B\times C)$$

and it is then obvious that the diagram

(1.20)
$$\begin{array}{ccc} (H_*B\otimes H_*C)\otimes(H_*B\otimes H_*C) & \xrightarrow{1\otimes\tau\otimes 1} & (H_*B\otimes H_*B)\otimes(H_*C\otimes H_*C) \\ & & \downarrow{\scriptstyle \mu_B\otimes\mu_C} \\ \downarrow\wr & & H_*B \otimes H_*C \\ & & \downarrow\wr \\ H_*(B\times C) & \xrightarrow{\mu_{B\times C}} & H_*(B\times C) \end{array}$$

is commutative. Here τ denotes the switching map for the tensor product of graded groups.

COROLLARY 1.5. *Let* A *be a torsion-free abelian group. Then* H_*A *is the exterior algebra* $E_Z A$ *over* A .

PROOF: First assume A finitely generated. We argue by induction on the number q of generators. Let q = 1 then the conclusion is trivial. For $q \geq 2$ we have $A = B \times C$ with the numbers of generators of B and C both smaller than q , so that by induction and Proposition 1.4

$$H_*A = H_*B \otimes H_*C = E_Z B \otimes E_Z C = E_Z(B \times C) = E_Z A . \tag{1.21}$$

For A non-finitely generated a direct limit argument easily yields the result.

PROPOSITION 1.6. *Let* C *be a finite cyclic group. Then the map* $\mu : H_*C \otimes H_*C \to H_*C$ *is trivial in positive dimensions.*

PROOF: Since $H_nC \neq 0$, for n = 0 and n odd, only, it suffices to consider

$$\mu : H_nC \otimes H_mC \to H_{n+m}C$$

for $n,m \geq 1$, odd. But then, n + m is even so that $H_{n+m}C = 0$, and the proposition is proved.

We finally note that we may, instead of taking integral homology, take homology with coefficients in a field K , the only change being that the tensor products be taken over K . Of course, the ring $H_*(N,K)$ associated with an abelian group N becomes a graded algebra over K .

PROPOSITION 1.7. *Let* K *be a field, and let* N_1,N_2 *be two abelian groups. Then*

$$H_*(N_1 \times N_2,K) \cong H_*(N_1,K) \otimes_K H_*(N_2,K) \tag{1.22}$$

as K-algebras.

PROOF: The proof is analogous to the proof of Proposition 1.4.

V.2. The Ganea Term

Let $N \rightarrowtail G \twoheadrightarrow Q$ be a central extension. We define the Ganea term

(2.1) $$\gamma : N \otimes G_{ab} \to H_2 G$$

as the restriction to $H_1 N \otimes H_1 G$ of $\mu : H_* N \otimes H_* G \to H_* G$. We shall first show how γ may be described in terms of a free presentation $R \rightarrowtail F \twoheadrightarrow G$. Recall that we have

(2.2) $$H_2 G \cong [F,F] \cap R/[F,R]$$

by Hopf's formula (see [43], p.204).

PROPOSITION 2.1. *If* $v \in F$ *is a representative of* $u \in N$ *and* $z \in F$ *is a representative of* $x \in G$, *then*

(2.3) $$\gamma(u \otimes x[G,G]) = [v,z][F,R] .$$

PROOF: By Proposition 1.1 the map $\gamma = \mu : H_1 N \otimes H_1 G \to H_2 G$ is induced by

(2.4) $$\mu([u]\otimes[x]) = [u|x]-[x|u] , u \in N , x \in G .$$

Now consider the short exact sequence

$$R_{ab} \overset{\kappa}{\rightarrowtail} ZG \otimes_F IF \overset{\nu}{\twoheadrightarrow} IG$$

associated with $R \rightarrowtail F \overset{f}{\twoheadrightarrow} G$ (see [43], p.198). If $s : G \to F$ is a function with $fs = 1_G$ and $s(e) = e$, then we may construct, as in the proof of Proposition II.4.4, a commutative diagram

$$(2.5)\qquad \begin{array}{ccccc} B_2' & \xrightarrow{\partial_2} & B_1' & \xrightarrow{\partial_1} & IG \\ \varphi_2\downarrow & & \varphi_1\downarrow & & \| \\ R_{ab} & \xrightarrow{\kappa} & ZG\otimes_F IF & \xrightarrow{\nu} & IG \end{array}$$

by setting

$$\varphi_1[x] = 1 \otimes (sx-1) \qquad , \quad x \in G ,$$

$$\varphi_2[x|y] = sxsy(s(xy))^{-1}[R,R] \quad , \quad x,y \in G .$$

Applying the functor $Z \otimes_G -$ to diagram (2.5) and using the fact that $H_2G = [F,F] \cap R/[F,R] = \ker(1 \otimes \kappa)$ we obtain

$$(2.6)\qquad \begin{aligned} \gamma(u \otimes x[G,G]) &= (\varphi_2([u|x]-[x|u])[F,R] \\ &= susx(s(ux))^{-1}(sxsu(s(xu))^{-1})^{-1}[F,R] \\ &= susx(su)^{-1}(sx)^{-1}[F,R] \end{aligned}$$

since $ux = xu$. Setting $su = v$, $sx = z$ the conclusion follows.

Note that we have also proved that the map γ as defined by (2.3) is independent of the chosen representatives and of the free presentation of G .

THEOREM 2.2. (Ganea [30]) *Let $E : N \xrightarrow{h} G \xrightarrow{g} Q$ be a central extension of groups. Then the sequence*

$$(2.7)\qquad N\otimes G_{ab} \xrightarrow{\gamma} H_2G \xrightarrow{g_*} H_2Q \xrightarrow{\delta_*^E} N \xrightarrow{h_*} G_{ab} \xrightarrow{g_*} Q_{ab} \to 0$$

is exact and natural.

PROOF: Only the exactness at H_2G remains to be proved. To this end let $R \rightarrowtail F \twoheadrightarrow G$ be a presentation of G and $S \rightarrowtail F \twoheadrightarrow Q$ the induced presentation of Q . Of course $R \rightarrowtail S \twoheadrightarrow N$ is a presentation of N . By Hopf's formula for H_2G , H_2Q we have

$$\begin{aligned}\ker(g_*:H_2G\to H_2Q) &= \ker([F,F]\cap R/[F,R]\to[F,F]\cap S/[F,S])\\ (2.8)\qquad &= R\cap[F,S]/[F,R]\\ &= [F,S]/[F,R]\ ,\end{aligned}$$

since $[F,S] \subseteq R$, in view of the fact that $N = S/R$ is central. On the other hand it follows from Proposition 2.1 that

$$(2.9)\qquad \operatorname{im}(\gamma:N\otimes G_{ab}\to H_2G) = [F,S]/[F,R]\ .$$

Thus Theorem 2.2 is proved.

We finally show that the map γ is closely related to the ordinary commutator in the group G . Let $N_1 \rightarrowtail G \twoheadrightarrow Q$ be a central extension, and let $N_2/N_1 \rightarrowtail G \twoheadrightarrow Q/N_2$ be a second central extension. Note that $N_2 \subseteq Z_2G$, the second center of G . Note also that $N_i = Z_iG$, $i=1,2$ are possible choices for N_1, N_2 . Consider then the associated sequences in homology.

$$(2.10)\qquad \begin{array}{ccccc} (N_2)_{ab}\otimes G_{ab} & & H_2G & & \\ \beta\downarrow & & \downarrow & & \\ N_2/N_1\otimes(G/N_1)_{ab} & \xrightarrow{\gamma} & H_2(G/N_1) & \to & H_2(G/N_2)\\ & & \delta_*\downarrow & & \\ & & N_1 & & \\ & & \downarrow & & \\ & & \vdots & & \end{array}$$

Here β denotes the obvious projection.

<u>PROPOSITION 2.3</u>. $\delta_*\gamma\beta(u[N_2,N_2]\otimes x[G,G]) = [u,x]$, $u\in N_2$, $x\in G$.

<u>PROOF</u>: Consider a free presentation $R\rightarrowtail F \twoheadrightarrow G$ of G and the associated presentations of G/N_i , $i = 1,2$

(2.11) $S \rightarrowtail F \twoheadrightarrow G/N_1 \; ; \; T \rightarrowtail F \twoheadrightarrow G/N_2$.

Note that $S/R \cong N_1$, $T/R \cong N_2$, $T/S \cong N_2/N_1$. Let $u \in N_2$, $x \in G$ and let $v \in T$, $z \in F$ be representatives of u , x respectively. We then have

$$\begin{aligned} \delta_* \gamma\beta(u[N_2,N_2] \otimes x[G,G]) &= \delta_* \gamma(vS \otimes z[F,F]S) \\ &= \delta_*([v,z][F,S]) \qquad (2.12) \\ &= [v,z]R \qquad \in S/R \cong N_1 . \end{aligned}$$

But in $F/R \cong G$ we have $[v,z]R = [u,x]$, so that our proof is complete.

Note that it follows from Proposition 2.3 that

(2.13) $[\ ,\ ] : Z_2G \times G \to Z_1G$

is a bilinear map. Of course, this is well-known and is easy to prove directly.

<u>REMARK</u>: If we consider an arbitrary variety $\underline{\underline{V}}$ instead of $\underline{\underline{Gr}}$ we obtain

<u>PROPOSITION 2.4</u>. <u>Let</u> $E : N \overset{h}{\rightarrowtail} G \overset{g}{\twoheadrightarrow} Q$ <u>be a central extension in</u> $\underline{\underline{V}}$. <u>Then the sequence</u>

(2.10) $N \otimes G_{ab} \xrightarrow{\gamma'} VG \xrightarrow{g_*} VQ \xrightarrow{\delta_*^E} N \xrightarrow{h_*} G_{ab} \xrightarrow{g_*} Q_{ab} \to 0$

<u>is exact and natural</u>.

<u>PROOF</u>: Let $f : F \twoheadrightarrow G$ be a $\underline{\underline{V}}$-free presentation of G and $gf : F \twoheadrightarrow Q$ the associated $\underline{\underline{V}}$-free presentation of Q . Then we obtain the following commutative diagram with exact rows and columns

$$
\begin{array}{ccccccccccccc}
 & & H_2F & = & H_2F & & & & & & & & \\
 & & \downarrow & & \downarrow & & & & & & & & \\
N\otimes G_{ab} & \xrightarrow{\gamma} & H_2G & \to & H_2Q & \to & N & \to & G_{ab} & \to & Q_{ab} & \to & 0 \\
 & \searrow^{\gamma'} & \downarrow & & \downarrow & & \| & & \| & & \| & & \\
 & & VG & \to & VQ & \to & N & \to & G_{ab} & \to & Q_{ab} & \to & 0 \\
 & & \downarrow & & \downarrow & & & & & & & & \\
 & & 0 & & 0 & & & & & & & &
\end{array}
$$

It is then obvious from Theorem 2.2 that the sequence (2.10) is also exact. Naturality follows from the naturality of sequence (2.7).

V.3. Various Classes of Central Extensions

In Sections 3, 5, 6 we shall develop what is commonly known as the theory of Schur [72], [73]. We shall do this in the variety $\underline{\underline{V}} = \underline{\underline{Gr}}$, mainly for notational reasons. Much of what we do, however, is true in any variety $\underline{\underline{V}}$ of exponent zero, if H_2- is replaced by V- and $H^2(-,-)$ is replaced by $\tilde{V}(-,-)$. Remarks indicating how to effect the translation are to be found at the end of Sections 5, 6.

Let Q be a fixed group. We first introduce the notion of isomorphic extensions of Q (see Gruenberg [35]). We say that two (not necessarily central) extensions $E : N \rightarrowtail G \twoheadrightarrow Q$ and $E' : N' \rightarrowtail G' \twoheadrightarrow Q$ are <u>isomorphic</u> if there is an isomorphism $f_1 : G \xrightarrow{\sim} G'$ inducing $f_2 : N \to N'$ such that the diagram

$$
(3.1) \qquad
\begin{array}{rccccc}
E : & N & \rightarrowtail & G & \twoheadrightarrow & Q \\
 & f_2\downarrow & & f_1\downarrow & & \| \\
E' : & N' & \rightarrowtail & G' & \twoheadrightarrow & Q
\end{array}
$$

is commutative. Accordingly, if Q is given, we may speak of the isomorphism class $[[E]]$ of an extension E . It is clear that equivalent extensions belong to the same isomorphism class, thus explaining our notation.

In the sequel we shall consider central extensions. We may then associate with the extension

(3.2) $$E : N \overset{h}{\rightarrowtail} G \overset{g}{\twoheadrightarrow} Q$$

the homomorphism $\Pi\Delta[E] : H_2Q \to N$ given by the universal coefficient sequence of $H^2(Q,N)$ (see II.5.1). By Proposition II.5.4 we have

(3.3) $$\Pi\Delta[E] = \delta_*^E ,$$

where δ_*^E is the "connecting map" in the 5-term homology sequence, associated with (3.2)

(3.4) $$H_2G \xrightarrow{g_*} H_2Q \xrightarrow{\delta_*^E} N \xrightarrow{h_*} G_{ab} \xrightarrow{g_*} Q_{ab} \to 0 .$$

We use properties of $\Pi\Delta[E]$ to define various classes of central extensions.

<u>DEFINITION</u>: The central extension (3.2) is called

(i) a <u>commutator extension</u>, if $\Pi\Delta[E] = 0$;

(ii) a <u>stem extension</u>, if $\Pi\Delta[E]$ is epimorphic;

(iii) a <u>stem cover</u>, if $\Pi\Delta[E]$ is isomorphic.

It is clear from naturality that the property of a central extension of Q of being in one of these classes depends only on its isomorphism class. Next we give various characterizations for a central extension to be in one of the classes (i), (ii), (iii). We first consider commutator extensions.

PROPOSITION 3.1. <u>For a central extension</u> $E : N \rightarrowtail^{h} G \xrightarrow{g} Q$ <u>the following statements are equivalent</u>.

(i) E <u>is a commutator extension</u>;

(ii) $N \rightarrowtail G_{ab} \twoheadrightarrow Q_{ab}$ <u>is exact</u>;

(iii) $g_* : [G,G] \xrightarrow{\sim} [Q,Q]$,

(iv) $N \cap [G,G] = e$.

<u>PROOF</u>: The equivalence of (i) and (ii) follows from the 5-term sequence (3.4), since $\Pi\Delta[E] = \delta_*^E$. Given (ii) it follows from the diagram

$$\begin{array}{ccccc} & & [G,G] & \xrightarrow{g_*} & [Q,Q] \\ & & \downarrow & & \downarrow \\ N & \rightarrowtail & G & \xrightarrow{g} & Q \\ \| & & \downarrow & & \downarrow \\ N & \rightarrowtail & G_{ab} & \xrightarrow{g_*} & Q_{ab} \end{array}$$

that $g_* : [G,G] \xrightarrow{\sim} [Q,Q]$. Thus (iii) holds. If (iii) holds then of course $N \cap [G,G] = e$. Finally if (iv) holds then $N \rightarrowtail G_{ab} \twoheadrightarrow Q_{ab}$ is exact.

PROPOSITION 3.2. <u>The equivalence classes of commutator extensions of</u> Q <u>are classified by</u> $\mathrm{Ext}(Q_{ab},N)$.

<u>PROOF</u>: The universal coefficient sequence

$$\mathrm{Ext}(Q_{ab},N) \overset{\Sigma}{\rightarrowtail} H^2(Q,N) \overset{\Pi}{\twoheadrightarrow} \mathrm{Hom}(H_2Q,N)$$

shows that E is a commutator extension, i.e. $\Pi\Delta[E] = 0$ if and only if $\Delta[E]$ lies in $\Sigma(\mathrm{Ext}(Q_{ab},N))$.

Note that if $[E']$ denotes the equivalence class of the abelian extension $N \rightarrowtail G_{ab} \twoheadrightarrow Q_{ab}$ then $\Sigma[E'] = \Delta[E]$. We now turn to stem extensions.

PROPOSITION 3.3. For a central extension $E : N \rightarrowtail^{h} G \overset{g}{\twoheadrightarrow} Q$ the following statements are equivalent.

(i) E is a stem extension;

(ii) $\delta_*^E : H_2Q \to N$ is epimorphic;

(iii) $h_* : N \to G_{ab}$ is the zero map;

(iv) $g_* : G_{ab} \xrightarrow{\sim} Q_{ab}$;

(v) $N \subseteq [G,G]$.

PROOF: The equivalence of (i) and (ii) follows from (3.3). The equivalence of (ii), (iii), (iv) is clear from the 5-term sequence (3.4). The equivalence of (iv) and (v) is trivial.

PROPOSITION 3.4. For a central extension $E : N \rightarrowtail^{h} G \overset{g}{\twoheadrightarrow} Q$ the following statements are equivalent.

(i) E is a stem cover;

(ii) $\delta_*^E : H_2Q \xrightarrow{\sim} N$;

(iii) $g_* : G_{ab} \xrightarrow{\sim} Q_{ab}$, and $g_* : H_2G \to H_2Q$ is the zero map.

PROOF: This easily follows from the 5-term sequence (3.4).

We will continue the discussion of stem extensions and of stem covers in the subsequent sections.

V.4. Indecomposables

Let

(4.1) $$E : N \rightarrowtail^{h} G \overset{g}{\twoheadrightarrow} Q$$

be a central extension. By Proposition 1.3 we know that H_*G is a graded module over the graded ring H_*N . We may thus ask what the relations are between the extension (4.1) and the module structure of H_*G . In this section we shall give some results in this direction.

We associate with the H_*N-module H_*G the graded group I defined by

(4.2) $$I_n = \operatorname{coker}(\mu : \bigoplus_{\substack{i+j=n \\ i\geq 1}} H_iN \otimes H_jG \to H_nG) .$$

An element of H_nG whose image is non-zero in I_n is called indecomposable, and an element whose image is zero is called decomposable.

We shall discuss I_* in low dimensions. In dimension $n = 1$ we have to consider

(4.3) $$\mu : H_1N \otimes H_oG \to H_1G .$$

By the explicit description of μ in (1.7) this agrees with $h_* : N \to G_{ab}$. We may thus state

PROPOSITION 4.1. $I_1 \cong Q_{ab}$.

In dimension $n = 2$ we have to consider

$$\mu : (H_2N \otimes H_oG) \oplus (H_1N \otimes H_1G) \to H_2G .$$

It is clear from the explicit description of μ in (1.7) that $\mu|H_2N \otimes H_oG$ agrees with $h_* : H_2N \to H_2G$. By definition $\mu|H_1N \otimes H_1G$ is the Ganea map γ . Applying the naturality of γ to the diagram

$$\begin{array}{ccccc} N & \rightarrowtail & N & \twoheadrightarrow & e \\ \| & & \downarrow h & & \downarrow \\ N & \overset{h}{\rightarrowtail} & G & \overset{g}{\twoheadrightarrow} & Q \end{array}$$

we obtain

$$(4.4)\qquad \begin{array}{ccccccc} N\otimes N & \xrightarrow{\gamma} & H_2N & \to & 0 & \to & \dots \\ {\scriptstyle 1\otimes h_*}\downarrow & & \downarrow{\scriptstyle h_*} & & \downarrow & & \\ N\otimes G_{ab} & \xrightarrow{\gamma} & H_2G & \to & H_2Q & \to & \dots \end{array}$$

From this we may infer that

$$(4.5)\qquad \mu(H_2N \otimes H_oG) \subseteq \mu(H_1N \otimes H_1G)\ .$$

Thus in order to compute I_2 it is enough to consider $\gamma : N\otimes G_{ab} \to H_2G$.

PROPOSITION 4.2. $I_2 \cong \ker(\Pi\Delta[E] : H_2Q \to N)$.

PROOF: The sequence (2.7) associated with the extension (4.1) immediately yields

$$(4.6)\qquad I_2 = \operatorname{coker}\gamma \cong \operatorname{im}(g_*:H_2G\to H_2Q) = \ker(\delta_*^E:H_2Q\to N)\ .$$

But by Proposition II.5.4 we have $\delta_*^E = \Pi\Delta[E]$.

COROLLARY 4.3. Let $E : N \rightarrowtail G \twoheadrightarrow Q$ be a central extension.

(i) E is a stem extension if and only if all elements in dimension 1 are indecomposable, i.e. $I_1 \cong G_{ab}$.

(ii) E is a stem cover if and only if all elements in dimension 1 are indecomposable, i.e. $I_1 \cong G_{ab}$ and all elements in dimension 2 are decomposable, i.e. $I_2 = 0$.

(iii) E is a commutator extension if and only if $I_2 \cong H_2Q$.

PROOF: This immediately follows from the definition of these extensions in Section 3 and Proposition 4.2.

We do not know of any far-reaching results in higher dimensions; however we may prove the following results which simplify the calculation of I_* . Note that Proposition 4.4(i) generalizes (4.5).

PROPOSITION 4.4. (i) For every $n \geq 2$ we have

$$\operatorname{im}(\mu : H_1N \otimes H_{n-1}G \to H_nG) \supseteq \operatorname{im}(\mu : H_2N \otimes H_{n-2}G \to H_nG) .$$

(ii) If N is torsion-free, then, for $2 \leq i \leq n$,

$$\operatorname{im}(\mu : H_1N \otimes H_{n-1}G \to H_nG) \supseteq \operatorname{im}(\mu : H_iN \otimes H_{n-i}G \to H_nG) .$$

PROOF: (i) By naturality the following square is commutative

$$\begin{array}{ccc} N \otimes N \otimes H_{n-2}G & \xrightarrow{\gamma \otimes 1} & H_2N \otimes H_{n-2}G \\ {\scriptstyle 1\otimes\mu}\downarrow & & \downarrow{\scriptstyle \mu} \\ N \otimes H_{n-1}G & \xrightarrow{\mu} & H_nG \end{array}$$

Since $\gamma : N \otimes N \to H_2N$ is surjective, $\gamma \otimes 1$ also is, whence the result.

(ii) We show that for $2 \leq i \leq n$,

$$\operatorname{im}(\mu : H_{i-1}N \otimes H_{n-i+1}G \to H_nG) \supseteq \operatorname{im}(\mu : H_iN \otimes H_{n-i}G \to H_nG) .$$

We have a commutative square

$$\begin{array}{ccc} H_{i-1}N \otimes N \otimes H_{n-i}G & \xrightarrow{\mu \otimes 1} & H_iN \otimes H_{n-i}G \\ {\scriptstyle 1\otimes\mu}\downarrow & & \downarrow{\scriptstyle \mu} \\ H_{i-1}N \otimes H_{n-i+1}G & \xrightarrow{\mu} & H_nG \end{array}$$

Since N is torsion-free, $H_*N = E_{\mathbb{Z}}N$ by Corollary 1.5. Thus $\mu : H_{i-1}N \otimes N \to H_iN$ is surjective and the result follows.

V.5. Stem Extensions and Stem Covers

The results in this and the next section constitute in principle what is known as the theory of Schur [72], [73]. It is because of the crucial role which the second homology group plays in this theory that H_2Q (or a group isomorphic to it) is often called the Schur multiplicator of Q. The terms stem extension and stem cover come from Hall's paper [37], see also [35].

PROPOSITION 5.1. Let U be a subgroup of H_2Q. Then there exists a stem extension E with $U = \ker \Pi\Delta[E]$.

PROOF: Let $N = H_2Q/U$. Choose any central extension $E : N \rightarrowtail G \twoheadrightarrow Q$ with $\Pi\Delta[E] : H_2Q \to N$ the canonical projection. The 5-term sequence associated with E then yields that $U = \ker \delta_*^E = \ker \Pi\Delta[E]$ and that E is a stem extension.

By definition of a stem cover, a stem extension is a stem cover if and only if $U = 0$. We remark that any stem cover $E : N \rightarrowtail G \twoheadrightarrow Q$ of Q is isomorphic to a stem cover $E' : H_2Q \rightarrowtail G' \twoheadrightarrow Q$ with $\Pi\Delta[E'] = 1_{H_2Q}$. To prove this let $\psi : N \to H_2Q$ be the inverse of $\Pi\Delta[E] : H_2Q \to N$ and consider the diagram

$$\begin{array}{ccc} H^2(Q,N) & \xrightarrow{\Pi} & \mathrm{Hom}(H_2Q,N) \\ \psi_* \downarrow & & \downarrow \psi_* \\ H^2(Q,H_2Q) & \xrightarrow{\Pi} & \mathrm{Hom}(H_2Q,H_2Q) \end{array} \qquad (5.1)$$

Then obviously $\Pi(\psi_*\Delta[E]) = 1_{H_2Q}$, so that we may choose E' in such a way that $\Delta[E'] = \psi_*\Delta[E]$. Proposition II.4.1 then yields the diagram

$$(5.2)\qquad \begin{array}{ccccccc} E: & N & \rightarrowtail & G & \twoheadrightarrow & Q \\ & \psi\downarrow & & \downarrow & & \| \\ E': & H_2Q & \rightarrow & G' & \twoheadrightarrow & Q \end{array}$$

showing that E and E' belong to the same isomorphism class.

The name stem cover is motivated by the following result.

<u>PROPOSITION 5.2</u>. <u>Every stem extension of Q is epimorphic image of some stem cover</u>.

<u>PROOF</u>: (G. Rinehart [69]) Let $E : N \rightarrowtail G \twoheadrightarrow Q$ be a stem extension, characterized by $\Delta[E] = \xi \in H^2(Q,N)$. Then $\varphi = \Pi(\xi) : H_2Q \to N$ is an epimorphism. Now consider the diagram

$$(5.3)\qquad \begin{array}{ccccc} \mathrm{Ext}(Q_{ab},H_2Q) & \overset{\Sigma}{\rightarrowtail} & H^2(Q,H_2Q) & \overset{\Pi}{\twoheadrightarrow} & \mathrm{Hom}(H_2Q,H_2Q) \\ \downarrow\varphi_* & & \downarrow\varphi_* & & \downarrow\varphi_* \\ \mathrm{Ext}(Q_{ab},N) & \overset{\Sigma}{\rightarrowtail} & H^2(Q,N) & \overset{\Pi}{\twoheadrightarrow} & \mathrm{Hom}(H_2Q,N) \end{array}$$

To prove our proposition we first find $\eta \in H^2(Q,H_2Q)$ with $\varphi_*(\eta) = \xi$ and $\Pi(\eta) = 1_{H_2Q}$. Let $\eta' \in H^2(Q,H_2Q)$ be such that $\Pi(\eta') = 1_{H_2Q}$. Then $\Pi(\xi-\varphi_*(\eta')) = \Pi(\xi)-\varphi_*\Pi(\eta') = \varphi-\varphi = 0$. Thus there exists $\mu \in \mathrm{Ext}(Q_{ab},N)$ with $\Sigma\mu = \xi-\varphi_*\eta'$. Now $\varphi_* : \mathrm{Ext}(Q_{ab},H_2Q) \to \mathrm{Ext}(Q_{ab},N)$ is epimorphic, since $\varphi : H_2Q \to N$ is epimorphic. It follows that there exists $\nu \in \mathrm{Ext}(Q_{ab},H_2Q)$ with $\varphi_*(\nu) = \mu$. Define $\eta = \Sigma(\nu)+\eta'$. Obviously η has the required properties. Now let $E' : H_2Q \rightarrowtail G' \twoheadrightarrow Q$ be an extension with $\Delta[E'] = \eta$. By Proposition II.4.1 we may then find $f : G' \to G$ with

$$(5.4)\qquad \begin{array}{ccccccc} E': & H_2Q & \rightarrowtail & G' & \twoheadrightarrow & Q \\ & \varphi\downarrow & & f\downarrow & & \| \\ E: & N & \rightarrowtail & G & \twoheadrightarrow & Q \end{array}$$

commutative. Since φ is surjective, f also is. Thus the proof is complete.

From the definition of stem covers we may easily infer

PROPOSITION 5.3. *There are as many different isomorphism classes of stem covers of* Q *as there are elements in* $\mathrm{Ext}(Q_{ab},H_2Q)$.

PROPOSITION 5.4. *There is only one isomorphism class of stem covers of* Q *if and only if* $\mathrm{Ext}(Q_{ab},H_2Q) = 0$.

In this case we shall speak of *the* stem cover of Q .

PROPOSITION 5.5. *Let* $\bar{E} : H_2\bar{Q} \rightarrowtail \bar{G} \twoheadrightarrow \bar{Q}$ *be a stem cover and let* $E : N \rightarrowtail G \twoheadrightarrow Q$ *be a stem extension. Then, if* $\mathrm{Ext}(\bar{Q}_{ab},N) = 0$, *every homomorphism* $f : \bar{Q} \to Q$ *can be lifted to a map* $f' : \bar{G} \to G$.

PROOF: (G. Rinehart [69]) Consider first the diagram

$$\begin{array}{ccccc} \mathrm{Ext}(Q_{ab},N) & \overset{\Sigma}{\rightarrowtail} & H^2(Q,N) & \overset{\Pi}{\twoheadrightarrow} & \mathrm{Hom}(H_2Q,N) \\ f^*\downarrow & & f^*\downarrow & & f^*\downarrow \\ \mathrm{Ext}(\bar{Q}_{ab},N) & \overset{\Sigma}{\rightarrowtail} & H^2(\bar{Q},N) & \overset{\Pi}{\twoheadrightarrow} & \mathrm{Hom}(H_2\bar{Q},N) \end{array} \tag{5.5}$$

Let $\Delta[E] = \xi \in H^2(Q,N)$. Then we may use $\varphi = f^*(\Pi(\xi)) : H_2\bar{Q} \to N$ to obtain the diagram

$$\begin{array}{ccccc} \mathrm{Ext}(\bar{Q}_{ab},H_2\bar{Q}) & \overset{\Sigma}{\rightarrowtail} & H^2(\bar{Q},H_2\bar{Q}) & \twoheadrightarrow & \mathrm{Hom}(H_2\bar{Q},H_2\bar{Q}) \\ \varphi_*\downarrow & & \varphi_*\downarrow & & \varphi_*\downarrow \\ \mathrm{Ext}(\bar{Q}_{ab},N) & \overset{\Sigma}{\rightarrowtail} & H^2(\bar{Q},N) & \twoheadrightarrow & \mathrm{Hom}(H_2\bar{Q},N) \end{array} \tag{5.6}$$

Since $\eta = \Delta[\bar{E}] \in H^2(\bar{Q},H_2\bar{Q})$ is a stem cover, we may suppose $\Pi(\eta) = 1_{H_2Q}$. (Otherwise replace $\bar{E}$ by an isomorphic extension with this property.) But then $\varphi_*\Pi(\eta) = \varphi$. Since $\mathrm{Ext}(\bar{Q}_{ab},N) = 0$, it is clear that $\varphi_*(\eta) = f^*(\xi)$. The conclusion then follows from Proposition II.4.3.

COROLLARY 5.6. (Schur [72]) *Let* $f : \bar{Q} \to PGL(n,\mathbb{C})$ *be a complex projective representation of* Q . *Then, if* $\bar{G}$ *is a stem cover of* $\bar{Q}$ *there exists a complex linear representation* $f' : \bar{G} \to GL(n,\mathbb{C})$ *such that* f' *induces* f .

PROOF: Consider the extension

$$\mathbb{C}^* \rightarrowtail GL(n,\mathbb{C}) \twoheadrightarrow PGL(n,\mathbb{C}) \quad . \tag{5.7}$$

Since $\mathbb{C}^*$ is injective (it contains all roots), $Ext(\bar{Q}_{ab},\mathbb{C}^*) = 0$ and we may apply our Proposition 5.5 to yield $f' : \bar{G} \to GL(n,\mathbb{C})$ with

$$\begin{array}{ccccc} H_2\bar{Q} & \rightarrowtail & \bar{G} & \twoheadrightarrow & \bar{Q} \\ \downarrow & & \downarrow f' & & \downarrow f \\ \mathbb{C}^* & \rightarrowtail & GL(n,\mathbb{C}) & \twoheadrightarrow & PGL(n,\mathbb{C}) \end{array} \tag{5.8}$$

commutative.

REMARK: We note that the proofs of Propositions 5.1 through 5.5 are purely formal; they only use the universal coefficient exact sequence for $H^2(Q,N)$, see (II.5.1). Since for any variety $\underline{\underline{V}}$ of exponent zero an analogous universal coefficient exact sequence holds (see (III.8.8)), we have results for $\underline{\underline{V}}$ corresponding to Propositions 5.1 through 5.5. We refrain from stating them explicitly since they are obtained automatically by replacing H_2- by $V-$ and $H^2(-,-)$ by $\tilde{V}(-,-)$ in Propositions 5.1 through 5.5.

The situation for varieties $\underline{\underline{V}}$ of exponent $q > 0$ seems to be more complicated. It is clear that one hopes to replace H_2- by V^q- and $H^2(-,-)$ by $\tilde{V}(-,-)$. However, there is no short exact universal coefficient sequence for $\tilde{V}(Q,N)$; in particular, Π in (III.8.1) will not be epimorphic, in general. Under the additional hypothesis on Q that $Q_{ab} = Q^q_{ab}$ be projective over Z/qZ we have (see (III.8.11))

$$\Pi : \tilde{V}(Q,N) \tilde{\to} \mathrm{Hom}(V^q Q,N)$$

so that in this case we immediately obtain results analogous to Propositions 5.1 through 5.5. Note that the hypothesis corresponding to $\mathrm{Ext}(\bar{Q}_{ab},N) = 0$ will automatically be satisfied, when $\bar{Q}_{ab}$ is Z/qZ-projective. Note finally that if q is square free, Q_{ab} will always be Z/qZ-projective.

V.6. Central Extensions of Perfect Groups

The theory of central extensions is particularily nice, when the quotient group Q is perfect, i.e. when $Q = [Q,Q]$. With this case in mind we shall prove a series of results in this section. Since for the particular results the hypothesis that Q be perfect is unnecessarily strong we shall usually use weaker hypotheses in the statements of our propositions.

PROPOSITION 6.1. *Suppose that* $E : N \overset{h}{\rightarrowtail} G \overset{g}{\twoheadrightarrow} Q$ *and* $E' : N' \overset{h'}{\rightarrowtail} G' \overset{g'}{\twoheadrightarrow} Q'$ *are two central extensions. Let* $r : N \to N'$ *and* $s : Q \to Q'$ *be group homomorphisms, and suppose that* $\mathrm{Ext}(Q_{ab},N') = 0$.

(i) *There exists* $t : G \to G'$ *inducing* r,s *if and only if*

$$\begin{array}{ccc} H_2Q & \xrightarrow{\delta_*^E} & N \\ s_*\downarrow & & \downarrow r \\ H_2Q' & \xrightarrow{\delta_*^{E'}} & N' \end{array} \qquad (6.1)$$

is commutative.

(ii) *If* t *exists, it is unique if and only if* $\mathrm{Hom}(Q_{ab},N') = 0$.

PROOF: If t exists, then clearly (6.1) is commutative. To prove the converse consider the diagram

$$\begin{array}{ccccc} \mathrm{Ext}(Q_{ab},N) & \overset{\Sigma}{\rightarrowtail} & H^2(Q,N) & \overset{\Pi}{\twoheadrightarrow} & \mathrm{Hom}(H_2Q,N) \\ \downarrow & & \downarrow r_* & & \downarrow r_* \\ 0 & \longrightarrow & H^2(Q,N') & \overset{\Pi''}{\underset{\sim}{\longrightarrow}} & \mathrm{Hom}(H_2Q,N') \\ \uparrow & & \uparrow s^* & & \uparrow s^* \\ \mathrm{Ext}(Q'_{ab},N') & \overset{\Sigma'}{\rightarrowtail} & H^2(Q',N') & \overset{\Pi'}{\twoheadrightarrow} & \mathrm{Hom}(H_2Q',N') \end{array}$$

Let $\xi = \Delta[E]$ and $\xi' = \Delta[E']$. Using Proposition II.5.4 and the commutativity of (6.1) we then obtain

$$r_*(\Pi(\xi)) = r\delta_*^E = \delta_*^{E'}s_* = s^*(\delta_*^{E'}) = s^*(\Pi'(\xi')) .$$

Since Π'' is an isomorphism this implies $r_*(\xi) = s^*(\xi')$, so that Proposition II.4.3 yields the existence of t .

(ii) If t exists, then

$$(6.2)\qquad \begin{array}{ccccc} N & \overset{h}{\rightarrowtail} & G & \overset{g}{\twoheadrightarrow} & Q \\ r\downarrow & & t\downarrow & & s\downarrow \\ N' & \overset{h'}{\rightarrowtail} & G' & \overset{g'}{\twoheadrightarrow} & Q' \end{array}$$

is commutative. Let $t' : G \to G'$ be another map making (6.2) commutative. Then $t' = t \cdot h'fg$ for some homomorphism $f : Q \to N'$. Conversely if $f : Q \to N'$ is a homomorphism then $t' = t \cdot h'fg$ makes diagram (6.2) commutative. Thus t is unique if and only if $\mathrm{Hom}(Q_{ab},N') = 0$.

COROLLARY 6.2. (Eckmann-Hilton-Stammbach [22]). Under the hypotheses of Proposition 6.1 with Q perfect the map $t : G \to G'$ exists and is unique if and only if diagram (6.1) is commutative.

We recall that there is only one isomorphism class of stem covers if $\mathrm{Ext}(Q_{ab}, H_2Q) = 0$ (see Proposition 5.3). In this case we are able to give a more or less complete description of the stem extensions of Q .

PROPOSITION 6.3. *Let* $\mathrm{Ext}(Q_{ab}, H_2Q) = 0$. *Then the isomorphism classes of stem extensions of* Q *are in one-to-one correspondence with the subgroups* U *of* H_2Q . *Moreover if* U *and* V *are two subgroups of* H_2Q , *then* $U \subseteq V$ *if and only if there is a map (necessarily surjective) from the stem extension corresponding to* U *to the stem extension corresponding to* V .

PROOF: Let $E : N \rightarrowtail G \twoheadrightarrow Q$ be a stem extension. Associate with E the subgroup $U = \ker(\Pi\Delta[E] : H_2Q \to N)$ of H_2Q . It is clear that isomorphic stem extensions yield the same subgroup of H_2Q .

Conversely let $U \subseteq H_2Q$ be given. Set $N = H_2Q/U$ and consider the canonical projection $\tau : H_2Q \to N$. Since $\mathrm{Ext}(Q_{ab}, H_2Q) = 0$ we have $\mathrm{Ext}(Q_{ab}, N) = 0$, so that the universal coefficient exact sequence for $H^2(Q,N)$ yields an isomorphism

$$\Pi : H^2(Q,N) \xrightarrow{\sim} \mathrm{Hom}(H_2Q, N) .$$

The map τ thus determines a unique equivalence class $[E]$ of extensions with $\Pi\Delta[E] = \tau$. Since $\Pi\Delta[E] = \delta_*^E$ is surjective, E is a stem extension and the subgroup of H_2Q associated with E is U . Finally if $E' : N' \rightarrowtail G' \twoheadrightarrow Q$ is another stem extension associated with U we may clearly find $r : N \to N'$ making the diagram

$$\begin{array}{ccccc} U & \rightarrowtail & H_2Q & \xrightarrow{\delta_*^{E}} & N \\ \| & & \| & & \downarrow r \\ U & \rightarrowtail & H_2Q & \xrightarrow{\delta_*^{E'}} & N' \end{array} \tag{6.3}$$

commutative. It follows that r is isomorphic, and Proposition 6.1

asserts the existence of $t : G \to G'$ inducing r . Hence E and E' lie in the same isomorphism class.

Now let the commutative diagram

$$\begin{array}{cccccc} E : & N & \rightarrowtail & G & \twoheadrightarrow & Q \\ (6.4) \qquad & r\downarrow & & t\downarrow & & \| \\ E' : & N' & \rightarrowtail & G' & \rightarrow & Q \end{array}$$

of stem extensions be given. Then

$$\begin{array}{ccc} H_2Q & \xrightarrow{\delta_*^E} & N \\ \| & & \downarrow r \\ H_2Q & \xrightarrow{\delta_*^{E'}} & N' \end{array}$$

is commutative, so that $U = \ker \delta_*^E \subseteq \ker \delta_*^{E'} = V$.

To prove the converse, we first recall that every stem extension is isomorphic to an extension E with $\Pi\Delta[E]$ the canonical projection. It is thus enough to consider those. Thus let $U \subseteq V \subseteq H_2Q$. Set $N = H_2Q/U$, $N' = H_2Q/V$. Then we have a commutative square of canonical projections

$$(6.5) \qquad \begin{array}{ccc} H_2Q & \xrightarrow{\tau} & N \\ \| & & \downarrow r \\ H_2Q & \xrightarrow{\sigma} & N' \end{array}$$

Now if $E : N \rightarrowtail G \twoheadrightarrow Q$ is an extension with $\Pi\Delta[E] = \delta_*^E = \tau$ and $E' : N' \rightarrowtail G' \twoheadrightarrow Q$ an extension with $\Pi\Delta[E'] = \delta_*^{E'} = \sigma$, then by Proposition 6.1 diagram (6.5) implies the existence of $t : G \to G'$ inducing r . Since r is surjective, t is also; and the proof is complete.

We remark that if Q is perfect it follows from Proposition 6.1 that the map t in (6.4) is uniquely determined by r .

PROPOSITION 6.4. *Let* Q *be perfect and let* $E : N \rightarrowtail G \twoheadrightarrow Q$ *be a stem extension. Then*

$$(6.6) \qquad 0 \to H_2G \to H_2Q \xrightarrow{\delta_*^E} N \to 0$$

is exact.

PROOF: This immediately follows from sequence (2.7) associated with E.

Note that it follows from Propositions 6.3 and 6.4 that if Q is perfect, the second homology groups of the stem extensions of Q are precisely the subgroups of H_2Q .

COROLLARY 6.5. *Let* Q *be perfect and let* $E : N \rightarrowtail G \twoheadrightarrow Q$ *be a central extension. Then* E *is the stem cover of* Q *if and only if* $H_1G = 0$ *and* $H_2G = 0$.

REMARK: It easily follows from Proposition 3.4 that if $E : N \rightarrowtail G \twoheadrightarrow Q$ is a stem extension with $H_2G = 0$ then it is a stem cover. Proposition 6.5 shows that the converse is true if, in addition, Q is perfect. In general, however, we may have stem covers with $H_2G \neq 0$. For example, let F be a non-commutative (absolutely) free group. Then it is easily seen that the central extension

$$(6.7) \qquad E : F_n^o/F_{n+1}^o \rightarrowtail F/F_{n+1}^o \twoheadrightarrow F/F_n^o \quad , \quad n \geq 2$$

is a stem cover of F/F_n^o . But we have

$$(6.8) \qquad H_2(F/F_{n+1}^o) = F_{n+1}^o/F_{n+2}^o \neq 0 .$$

The following proposition gives, for Q perfect, a description of the stem cover of Q in terms of a free presentation of Q .

PROPOSITION 6.6. Let Q be perfect, and let

$$S \rightarrowtail F \xrightarrow{f} Q$$

be a free presentation. Then

$$H_2Q \rightarrowtail [F,F]/[F,S] \xrightarrow{g} Q \qquad (6.9)$$

is the stem cover of Q, the map g being induced by f.

PROOF: Since Q is perfect we have $[F,F]S = F$ and therefore

$$[F,F]/[F,F]\cap S \cong [F,F]S/S \cong Q ,$$

the isomorphism being induced by f. We may thus consider the central extension

$$\begin{array}{lcccccc} E : & [F,F]\cap S/[F,S] & \rightarrowtail & [F,F]/[F,S] & \twoheadrightarrow & [F,F]/[F,F]\cap S \\ & \| & & \| & & \wr\| \\ E : & H_2Q & \rightarrowtail & G & \twoheadrightarrow & Q \end{array}$$

It is easy to see that it is characterized by $\Pi\Delta[E] = \delta_*^E = 1_{H_2Q}$. Thus it is the stem cover.

The following proposition is a result of Eckmann-Hilton-Stammbach [22].

PROPOSITION 6.7. Let $E : N \overset{h}{\rightarrowtail} G \overset{g}{\twoheadrightarrow} Q$ be a central extension and let $f : X \to Q$ be a homomorphism with X perfect. Then there exists $k : X \to G$ with $f = gk$ if and only if

$$f_*(H_2X) \subseteq g_*(H_2G) . \qquad (6.10)$$

If k exists, it is uniquely determined.

PROOF: If k exists, then clearly (6.10) holds. To prove the converse set $Q' = \operatorname{im} f \subseteq Q$ and $S = \ker f$. We may then consider the diagram

$$\text{(6.11)} \qquad \begin{array}{ccccc} S & \rightarrowtail & X & \twoheadrightarrow & Q' \\ \downarrow & & \downarrow & & \| \\ S/[X,S] & \rightarrowtail & X/[X,S] & \twoheadrightarrow & Q' \end{array}$$

Set, for short, $S' = S/[X,S]$, $X' = X/[X,S]$. Also, denote by $f' : X' \to Q$ the map induced by $f : X \to Q$. The diagram (6.11) gives rise to the diagram of 5-term sequences

$$\begin{array}{ccccccc} H_2X & \to & H_2Q' & \to & S/[X,S] & \to & \dots \\ \downarrow & & \| & & \| & & \\ H_2X' & \to & H_2Q' & \to & S' & \to & \dots \end{array}$$

whence immediately $f'_*H_2X' = f_*H_2X \subseteq g_*H_2G \subseteq H_2Q$.

It remains to construct $k' : X' \to G$ such that $f' = gk'$, i.e. such that the diagram

$$\text{(6.12)} \qquad \begin{array}{ccccc} S' & \rightarrowtail & X' & \twoheadrightarrow & Q' \\ {\scriptstyle s'}\downarrow & & {\scriptstyle k'}\downarrow & & \downarrow \\ N & \rightarrowtail & G & \overset{g}{\twoheadrightarrow} & Q \end{array}$$

is commutative. To this end consider

$$\text{(6.13)} \qquad \begin{array}{ccccccccccc} & & H_2X' & \to & H_2Q' & \to & S' & \to & 0 & & \\ & & & & \downarrow & & & & & & \\ & & H_2G & \xrightarrow{g_*} & H_2Q & \to & N & \to & G_{ab} & \to & Q_{ab} \to 0 \end{array}$$

Since $f'_*H_2X' \subseteq g_*H_2G$ we obtain a unique map $s' : S' \to N$ making (6.13) commutative. Since X is perfect, it follows that Q' is perfect. Hence $\operatorname{Ext}(Q'_{ab},N) = 0 = \operatorname{Hom}(Q'_{ab},N)$. Thus Proposition 6.1 yields the existence of a uniquely determined $k' : X' \to G$ making (6.12) commutative, completing the proof of Proposition 6.7.

The reader may compare this result with well-known theorems in covering space theory for connected topological spaces. It then becomes apparent that the H_2 plays here a role analogous to π_1 in that theory.

REMARK: The proofs of this sections are purely formal, the main tools being the 5-term homology sequence and the universal coefficient exact sequence for $H^2(-,-)$. It is thus obvious that results analogous to those stated hold in any variety $\underline{\underline{V}}$ of exponent zero if H_2- is replaced by V- and $H^2(-,-)$ is replaced by $\tilde{V}(-,-)$. We leave the obvious translation of the statements to the reader.

V.7. A Generalization of the Theory of Schur

In the next two sections we shall present ideas due to Evens [28]. In Section 7 we shall generalize the theory of Schur, and in Section 8 we shall apply the theory to group-theoretical problems.

The generalization of the theory of Schur consists basically in substituting for the integral homology the homology with coefficients in Z/qZ, where q is any positive integer. The lower central series will then be replaced by the lower central(q) series $\{G_n^q\}$ as defined in (I.1.2). The upper central series (I.1.4) will have to be replaced by the upper central(q) series $\{Z_n^q\}$. It is defined as follows. Let

(7.1) $$Z^qG = \{x \in ZG \mid x^q = e\} .$$

Then the terms Z_n^q are defined recursively by

(7.2) $$Z_o^qG = e \;,\; Z_n^qG/Z_{n-1}^qG = Z^q(G/Z_{n-1}^qG) .$$

We call an extension

(7.3) $$E : N \overset{h}{\rightarrowtail} G \overset{g}{\twoheadrightarrow} Q$$

central(q), if $N \subseteq Z^qG$. Note that then N is a Z/qZ-module. It follows from (II.3.13) that (7.3) gives rise to a 5-term sequence

(7.4) $$H_2^qG \xrightarrow{g_*} H_2^qQ \xrightarrow{\delta_*^E} N \xrightarrow{h_*} G_{ab}^q \xrightarrow{g_*} Q_{ab}^q \rightarrow 0 .$$

From (II.5.3) we recall the universal coefficient exact sequence

(7.5) $$0 \rightarrow \mathrm{Ext}^1_{Z/qZ}(Q_{ab}^q, N) \overset{\Sigma}{\rightarrow} H^2(Q,N) \overset{\Pi}{\rightarrow} \mathrm{Hom}_{Z/qZ}(H_2^qQ, N) \rightarrow \ldots$$

Thus, as in Section 3, we may associate with the central(q) extension (7.3) a Z/qZ-homomorphism

(7.6) $$\Pi\Delta[E] : H_2^qQ \rightarrow N .$$

By Proposition II.5.4 we may identify $\Pi\Delta[E]$ with the homomorphism δ_*^E in (7.4). Proceeding as in Section 3 we may define various classes of central(q) extensions using properties of $\Pi\Delta[E]$.

<u>DEFINITION</u>: The central(q) extension (7.3) is called

(i) a <u>commutator</u>(q) <u>extension</u>, if $\Pi\Delta[E] = 0$;

(ii) a <u>stem</u>(q) <u>extension</u>, if $\Pi\Delta[E]$ is epimorphic;

(iii) a <u>stem</u>(q) <u>cover</u>, if $\Pi\Delta[E]$ is isomorphic.

Since $\Pi\Delta[E] = \delta_*^E$, sequence (7.4) then immediately yields the following propositions that correspond to Propositions 3.1 , 3.3 , 3.4.

<u>PROPOSITION 7.1</u>. <u>For the central</u>(q) <u>extension</u> (7.3) <u>the following statements are equivalent</u>.

(i) E <u>is a commutator</u>(q) <u>extension</u>;

(ii) $N \rightarrowtail G_{ab}^q \twoheadrightarrow Q_{ab}^q$ <u>is exact</u>;

(iii) $g_* : G \#_q G \rightarrow Q \#_q Q$;

(iv) $N \cap G \#_q G = e$.

PROPOSITION 7.2. For the central(q) extension (7.3) the following statements are equivalent.

(i) E is a stem(q) extension;

(ii) $\delta_*^E : H_2^q Q \to N$ is epimorphic;

(iii) $h_* : N \to G_{ab}^q$ is the zero map;

(iv) $g_* : G_{ab}^q \cong Q_{ab}^q$;

(v) $N \subseteq G \#_q G$.

PROPOSITION 7.3. For the central(q) extension (7.3) the following statements are equivalent.

(i) E is a stem(q) cover;

(ii) $\delta_*^E : H_2^q Q \cong N$;

(iii) $g_* : G_{ab}^q \cong Q_{ab}^q$, and $g_* : H_2^q G \to H_2^q Q$ is the zero map.

In general the statements analogous to Propositions 5.1 , 5.2 , 5.3 , 5.4 do not seem to be true. However, if Q_{ab}^q is Z/qZ-projective, for example if q is square free, then we again obtain meaningful results. For then we have an isomorphism

$$\Pi : H^2(Q,N) \cong \mathrm{Hom}_{Z/qZ}(H_2^q Q,N) .$$

Proceeding analogously to the proof of Proposition 5.1 we obtain

PROPOSITION 7.4. Suppose Q_{ab}^q is Z/qZ-projective. Let $U \subseteq H_2^q Q$. Then there exists a stem(q) extension E with $\ker \Pi\Delta[E] = U$.

An argument analogous to the one used in the proof of Proposition 5.2 establishes

PROPOSITION 7.5. If Q_{ab}^q is Z/qZ-projective, there is precisely one isomorphism class of stem(q) covers of Q . Every stem(q) extension of Q is an epimorphic image of that stem(q) cover.

PROPOSITION 7.6. *Let* $E' : H_2^q Q' \rightarrowtail G' \twoheadrightarrow Q'$ *be a stem*(q) *cover and let* $E : N \rightarrowtail G \twoheadrightarrow Q$ *be a stem*(q) *extension. Then, if* $\mathrm{Ext}^1_{Z/qZ}(Q'^q_{ab}, N) = 0$, *every homomorphism* $f : Q' \to Q$ *can be lifted to a map* $f' : G' \to G$.

The proof is analogous to the proof of Proposition 5.5.

PROPOSITION 7.7. *Let* Q^q_{ab} *be a projective* Z/qZ-*module. Then the isomorphism class of stem*(q) *extensions of* Q *are in one-to-one correspondence with the subgroups* U *of* $H_2^q Q$. *Moreover, if* U *and* V *are two subgroups of* $H_2^q Q$, *then* $U \subseteq V$ *if and only if there is a map (necessarily surjective) from the stem*(q) *extension corresponding to* U *to the stem*(q) *extension corresponding to* V .

The proof is analogous to the proof of Proposition 6.3.

V.8. Terminal and Unicentral Groups

We recall from Section 7 the definition of a stem(q) extension. The extension

(8.1) $$E : N \overset{h}{\rightarrowtail} G \overset{g}{\twoheadrightarrow} Q$$

is called a *stem*(q) *extension* if

(8.2) $$N \subseteq Z^q G \cap G_2^q ,$$

where q is any positive integer. If we use the term stem(0) extension to denote stem extensions then we may include this case by allowing $q = 0$, also. We shall generalize the notion of a stem(q) extension as follows

DEFINITION: The extension (8.1) is called an m-*stem*(q) *extension*, $m \geq 1$ if

(8.3) $$N \subseteq Z^qG \cap G^q_{m+1} .$$

Note that a 1-stem(q) extension is just a stem(q) extension. Note also that if (8.1) is an m-stem(q) extension, then g induces isomorphisms

(8.4) $$g_k : G/G^q_k \xrightarrow{\sim} Q/Q^q_k \quad , \quad k = 1,\ldots,m+1 .$$

The following proposition yields a homological characterization of m-stem(q) extensions.

<u>PROPOSITION 8.1</u>. <u>A stem</u>(q) <u>extension</u> $E : N \rightarrowtail G \twoheadrightarrow Q$ <u>is an</u> m-<u>stem</u>(q) <u>extension if and only if</u>

(8.5) $$\ker \delta^E_* + \ker \tau_m = H^q_2Q ,$$

<u>where</u> $\delta^E_* = \Pi\Delta[E] : H_2Q \to N$ <u>and</u> $\tau_m : H^q_2Q \to H^q_2(Q/Q^q_m)$.

<u>PROOF</u>: Consider the diagram

$$\begin{array}{ccccc} N & \rightarrowtail & G & \twoheadrightarrow & Q \\ \downarrow & & \| & & \downarrow \\ NG^q_m & \rightarrowtail & G & \twoheadrightarrow & Q/Q^q_m \end{array}$$

and the associated 5-term sequences

$$\begin{array}{ccccccccccc} & & \ker \tau_m & \overset{\bar{\delta}}{\dashrightarrow} & N \cap G^q_{m+1} & & & & & & \\ & & \downarrow & & \downarrow & & & & & & \\ H^q_2G & \to & H^q_2Q & \xrightarrow{\delta^E_*} & N & \to & G^q_{ab} & \to & Q^q_{ab} & \to & 0 \\ \| & & \tau_m\downarrow & & \downarrow\nu & & \| & & \| & & \\ H^q_2G & \to & H^q_2(Q/Q^q_m) & \to & NG^q_m/G^q_{m+1} & \to & G^q_{ab} & \to & Q^q_{ab} & \to & 0 \end{array} \quad (8.6)$$

A simple diagram chase shows that $\bar{\delta}$ is epimorphic. Now if E is an m-stem(q) extension, i.e. if $N \subseteq G^q_{m+1}$, then $N \cap G^q_{m+1} = N$ and (8.5) follows. Conversely, if (8.5) holds, then ν must be the zero map, so that $N \subseteq G^q_{m+1}$. Thus the proof is complete.

PROPOSITION 8.2. Suppose there is no subgroup U of H_2Q with $U \neq H_2^qQ$ and $U + \ker \tau_m = H_2^qQ$. Then there is no non-trivial m-stem(q) extension of Q. Moreover, if $q = 0$ or if Q_{ab}^q is Z/qZ-projective the converse is also true.

PROOF: Of course, if we are given a non-trivial m-stem(q) extension of Q, then $U = \ker \delta_*^E$ is a subgroup of H_2^qQ with $U \neq H_2^qQ$ and $U + \ker \tau_m = H_2^qQ$.

Now let $q = 0$ or Q_{ab}^q be Z/qZ-projective. Given $U \subseteq H_2^qQ$ with $U + \ker \tau_m = H_2^qQ$, Proposition 7.4 enables us to find a stem(q) extension E with $\ker \delta_*^E = \ker \Pi\Delta[E] = U$. By Proposition 8.1 it is an m-stem(q) extension. It is non-trivial if $U \neq H_2Q$.

COROLLARY 8.3. If $\ker \tau_m = 0$, then there are no non-trivial m-stem(q) extensions of Q.

COROLLARY 8.4. Let Q be a finite p-group. Suppose $q = 0$ or $q = p^k$. If $\ker \tau_m \subseteq p\, H_2^qQ$, there are no non-trivial m-stem(q) extensions of Q. Moreover if $q = 0$ or if Q_{ab}^q is Z/qZ-projective the converse is also true.

PROOF: Since Q is a finite p-group, so is H_2^qQ. If E is an m-stem(q) extension then $U = \ker \Pi\Delta[E]$ is a subgroup of H_2^qQ with $U + \ker \tau_m = H_2^qQ$. If we had $\ker \tau_m \subseteq p\, H_2^qQ$, then U would generate H_2^qQ mod pH_2^qQ, hence it necessarily would be the whole of H_2^qQ. Conversely, if $\ker \tau_m \not\subseteq pH_2^qQ$ we may find a subgroup $U \subseteq H_2^qQ$ with $U \neq H_2^qQ$ and $U + \ker \tau_m = H_2^qQ$. If $q = 0$ or if Q_{ab}^q is Z/qZ-projective Proposition 8.2 yields the existence of a non-trivial m-stem(q) extension of Q.

REMARK: Note that if $q = p^k$, the hypothesis of Corollary 8.4 may be weakened. It would be enough to suppose that Q is any group for which $H_2^q Q$ is finitely generated. This is always so if Q is finitely presentable. For $q = 0$ such a generalization does not seem to be possible.

DEFINITION: A nilpotent(q) group Q of class m is called terminal(q) if there are no non-trivial m-stem(q) extensions of Q.

Note that if Q is terminal(q) and G is nilpotent(q) then $G/G_{m+1}^q \cong Q$ implies $G_{m+1}^q = e$. Note also that if Q_1 and Q_2 are terminal(q), then their direct product $Q_1 \times Q_2$ is terminal(q), too.

PROPOSITION 8.5. *Let Q be nilpotent(q) of class m. If $\tau_m : H_2^q Q \to H_2^q(Q/Q_m^q)$ is monomorphic, then Q is terminal(q).*

PROOF: By Corollary 8.3 there are no non-trivial m-stem(q) extensions of Q.

PROPOSITION 8.6. *Let Q be a finite p-group. Suppose $q = 0$ or $q = p^k$. Then Q is nilpotent(q), of class m, say. If $\ker \tau_m \subseteq pH_2^q Q$, it is terminal(q). Moreover, if $q = 0$ or if Q_{ab}^q is Z/qZ-projective, the converse is also true.*

PROOF: By Lemma I.1.1 Q is nilpotent(q). The rest is an immediate consequence of Corollary 8.4.

DEFINITION: A group Q is called unicentral(q) if for every extension $N \rightarrowtail G \twoheadrightarrow Q$ with $N \subseteq Z^q G$ the sequence

$$(8.7) \qquad N \rightarrowtail Z^q G \twoheadrightarrow Z^q Q$$

is exact.

Our next proposition is an improvement due to Meier [62] of a result of Evens [28].

PROPOSITION 8.7. *Let* Q *be nilpotent*(q). *If*

$$(8.8) \qquad \sigma : H_2^q Q \to H_2^q(Q/Z^q Q)$$

is monomorphic then Q *is unicentral*(q). *Moreover, if* $q = 0$ *or if* Q_{ab}^q *is* Z/qZ-*projective, the converse is also true*.

PROOF: Let $E : N \overset{h}{\rightarrowtail} G \overset{g}{\twoheadrightarrow} Q$ be a central(q) extension and let $M = g^{-1}(Z^q Q)$. Since $M \supseteq N$, we obtain the diagram

$$(8.9) \qquad \begin{array}{ccccccccccc} & & 0 & & 0 & & & & & & \\ & & \downarrow & & \downarrow & & & & & & \\ & & \ker \sigma & \xrightarrow{\bar{\delta}} & N \cap (G \#_q M) & & & & & & \\ & & \downarrow & & \downarrow & & & & & & \\ H_2^q G & \to & H_2^q Q & \xrightarrow{\delta_*^E} & N & \to & G_{ab}^q & \to & Q_{ab}^q & \to & 0 \\ \| & & \sigma\downarrow & & \downarrow & & \| & & \downarrow & & \\ H_2^q G & \to & H_2^q(Q/Z^q Q) & \to & M/G \#_q M & \to & G_{ab}^q & \to & (Q/Z^q Q)_{ab}^q & \to & 0 \end{array}$$

It is easy to see that $\bar{\delta}$ is epimorphic. Thus if $\ker \sigma = 0$ it follows that $N \cap (G \#_q M) = e$. But, by the construction of M , we have $G \#_q M \subseteq N$, whence we conclude that $G \#_q M = e$, i.e. $M \subseteq Z^q G$. Since trivially $Z^q G \subseteq M$, the sequence

$$N \rightarrowtail Z^q G \twoheadrightarrow Z^q Q$$

is exact, so that Q is unicentral(q).

To prove the second part of the proposition we recall that if $q = 0$ or if Q_{ab}^q is Z/qZ-projective then there exist stem(q) covers of Q . Let

$$(8.10) \qquad E : H_2^q Q \rightarrowtail G \twoheadrightarrow Q$$

be such a stem(q) cover, so that δ_*^E is an isomorphism. Consider the diagram with exact rows and columns

$$\begin{array}{ccccccc}
 & & 0 & & 0 & & \\
 & & \downarrow & & \downarrow & & \\
 & & \ker \sigma' & \overset{\rho}{\to} & \ker \sigma & \overset{\bar{\delta}}{\to} & H_2^q Q \\
 & & \downarrow & & \downarrow & & \Vert \\
 & & H_2^q G & \overset{\rho}{\to} & H_2^q Q & \xrightarrow{\delta_*^E} & H_2^q Q \to 0 \\
 & & \sigma' \downarrow & & \sigma \downarrow & & \\
(8.11) & & H_2^q(G/Z^qG) & \overset{\rho}{\to} & H_2^q(Q/Z^qQ) & & \\
 & & \downarrow & & \downarrow & & \\
 & & Z^qG \cap G_2^q & \to & Z^qQ \cap Q_2^q & & \\
 & & \downarrow & & \downarrow & & \\
 & & 0 & & 0 & &
\end{array}$$

If Q is unicentral(q), then $G/Z^qG \cong Q/Z^qQ$ so that ρ is an isomorphism. The ker - coker sequence applied to (8.11) then yields

$$\ker \sigma' \overset{\rho}{\to} \ker \sigma \overset{\bar{\delta}}{\to} H_2^q Q \overset{\omega}{\to} Z^qG \cap G_2^q \to Z^qQ \cap Q_2^q \to 0$$

Since E is a stem(q) extension, the map ω is monomorphic. Thus $\bar{\delta}$ is the zero map, and hence $\ker \sigma = 0$.

COROLLARY 8.8. *Let Q be nilpotent(q). Suppose $q = 0$ or Q_{ab}^q is Z/qZ-projective. Then, if Q is unicentral(q), it is terminal(q).*

PROOF: Let Q be nilpotent(q) of class m . Since $Q_m^q \subseteq Z^qQ$ we have

$$\begin{array}{cccc}
 & H_2^q Q & \xrightarrow{\tau_m} & H_2^q(Q/Q_m^q) \\
(8.12) & & \searrow & \downarrow \\
 & & \sigma & H_2^q(Q/Z^qQ)
\end{array}$$

If Q is unicentral(q), then $\ker \sigma = 0$ by Proposition 8.7. Hence $\ker \tau_m = 0$, so that Q is terminal(q) by Proposition 8.5.

Note that if Q_1 , Q_2 are unicentral(q), their direct product Q need not be. To see this, let $q = 0$. Then there is a natural direct sum decomposition

$$H_2Q = H_2Q_1 \oplus H_2Q_2 \oplus Q_{1ab} \otimes Q_{2ab} \ ,$$

$$H_2(Q/ZQ) = H_2(Q_1/ZQ_1) \oplus H_2(Q_2/ZQ_2) \oplus (Q_1/ZQ_1)_{ab} \otimes (Q_2/ZQ_2)_{ab} \ .$$

Thus, in order to study σ we may study its components. It is then clear that

$$Q_{1ab} \otimes Q_{2ab} \to (Q_1/ZQ_1)_{ab} \otimes (Q_2/ZQ_2)_{ab}$$

need not be monomorphic. So, although $Q = Q_1 \times Q_2$ is terminal(q), it need not be unicentral(q).

V.9. On the Order of the Schur Multiplicator

In this section we shall give two estimates on the order of the Schur multiplicator of a finite nilpotent group G and an estimate on the rank of H_2G for a nilpotent group G.

PROPOSITION 9.1. (Green [33]) *Let* $|G| = p^m$. *Then*

(9.1) $$|H_2G| \leq p^{\frac{1}{2}m(m-1)} \ .$$

PROOF: We argue by induction on m. If $m = 0$, then $G = e$ and $H_2G = 0$; thus the assertion is true in this case. For $m \geq 1$ we may find a central subgroup N of G of order p. Denote by Q the corresponding factor group G/N. Then naturality of the sequence (2.10) applied to

$$\begin{array}{ccccc} N & \rightarrowtail & N & \twoheadrightarrow & e \\ \| & & \downarrow h & & \downarrow \\ N & \overset{h}{\rightarrowtail} & G & \overset{g}{\twoheadrightarrow} & Q \end{array}$$

yields the diagram

$$
(9.2) \qquad
\begin{array}{ccccccccccccc}
N \otimes N & \xrightarrow{\gamma'} & H_2N & & & & & & & & & & \\
{\scriptstyle 1 \otimes h_*}\downarrow & & \downarrow{\scriptstyle h_*} & & & & & & & & & & \\
N \otimes G_{ab} & \xrightarrow{\gamma} & H_2G & \to & H_2Q & \to & N & \to & G_{ab} & \to & Q_{ab} & \to & 0 \\
{\scriptstyle 1 \otimes g_*}\downarrow & \nearrow{\scriptstyle \gamma''} & & & & & & & & & & & \\
N \otimes Q_{ab} & & & & & & & & & & & & \\
\downarrow & & & & & & & & & & & & \\
0 & & & & & & & & & & & &
\end{array}
$$

Since N is cyclic, $H_2N = 0$. Hence there exists γ'' with $\gamma''(1 \otimes g_*) = \gamma$. Using the fact that $|Q| = p^{m-1}$ and applying the induction hypothesis we obtain

$$|H_2G| \leq |Q_{ab}| \cdot |H_2Q| \leq p^{m-1} \cdot p^{\frac{1}{2}(m-1)(m-2)} = p^{\frac{1}{2}m(m-1)}$$

and the proof is complete.

<u>COROLLARY 9.2</u>. <u>Let</u> G <u>be a nilpotent group of order</u> $n = p_1^{m_1} \cdot p_2^{m_2} \cdots p_\ell^{m_\ell}$. <u>Then</u>

$$(9.3) \qquad |H_2G| \leq \prod_{i=1}^{\ell} p_i^{\frac{1}{2} m_i (m_i - 1)} .$$

<u>PROOF</u>: Since G is nilpotent it is the direct product of its p_i-Sylow subgroups P_i, $i = 1, \ldots, \ell$. The Künneth-sequence (II.5.13) then yields

$$H_2G \cong \bigoplus_{i=1}^{\ell} H_2P_i$$

whence the assertion by Proposition 9.1.

We observe that the estimate (9.1) is best possible. For if $G = C_p \times \ldots \times C_p$, m-times, then $|G| = p^m$ and the Künneth-sequence (II.5.13) easily yields

$$|H_2G| = p^{\frac{1}{2} m(m-1)}$$

by induction on m . However, if we allow the structure of G to enter into consideration, then better estimates are clearly possible. We will state one of these in Proposition 9.5 , but first need some preliminaries.

Let G be nilpotent of class n ; then $G_n^o \neq e$, $G_{n+1}^o = e$, $Z_{n-1}G \neq G$ $Z_nG = G$. Consider then the central extension

(9.4) $$E : G_n^o \rightarrowtail G \twoheadrightarrow G/G_n^o .$$

Since $G/Z_{n-1}G$ is abelian, $G_2^o \subseteq Z_{n-1}G$ and the embedding of $Z_{n-1}G$ in G induces

$$\iota_* : G_n^o \otimes Z_{n-1}G/G_2^o \to G_n \otimes G_{ab} .$$

We may then prove

<u>PROPOSITION 9.3</u>. <u>Let</u> $\gamma : G_n^o \otimes G_{ab} \to H_2G$ <u>be the Ganea term</u> (2.1) <u>of the extension</u> (9.4). <u>Then</u>

(9.5) $$\gamma \circ \iota_* : G_n \otimes Z_{n-1}G/G_2^o \to H_2G$$

<u>is the zero map</u>.

<u>PROOF</u>: Consider a stem cover $E' : H_2G \overset{h}{\rightarrowtail} K \overset{g}{\twoheadrightarrow} G$ with $\Pi\Delta[E'] = 1$. Setting $H_2G = N$ we have

(i) $g(N) = e$,

(ii) $g^{-1}(G_n^o) = K_n^o \cdot N \subseteq Z_2K$,

(iii) $g(Z_nK) \supseteq Z_{n-1}G$.

Statements (i), (ii) are trivial. To prove (iii) we argue by induction to prove $g(Z_kK) \supseteq Z_{k-1}G$ for $k = 1,\ldots,n$. Clearly $g(Z_1K) \supseteq Z_oG=e$. Inductively we may assume $g(Z_{k-1}K) \supseteq Z_{k-2}G$. Set $M = g^{-1}(Z_{k-2}G)$, so that $M \subseteq Z_{k-1}K$. Thus we have

$$Z_k K/M \supseteq Z(K/M) \overset{g}{\to} Z(G/Z_{k-2}G) = Z_{k-1}G/Z_{k-2}G$$

whence $g(Z_k K) \supseteq Z_{k-1}G$. Next consider

$$\begin{array}{ccccc} (K_n^o N)_{ab} \otimes K_{ab} & & H_2K & & \\ \beta\downarrow & & \downarrow g_*=0 & & \\ G_n^o \otimes G_{ab} & \overset{\gamma}{\to} & H_2G & \to & \cdots \\ & & \downarrow \delta_*=1 & & \\ & & H_2G & & \\ & & \downarrow & & \\ & & \vdots & & \end{array} \tag{9.6}$$

Recall from Proposition 2.3 that $\delta_*\gamma\beta$ is the commutator in K . But since (see [15], p.3)

$$[K_n^o, Z_n K] = e$$

it follows that $[K_n^o N, Z_n K] = e$, and using (iii) we may infer that

$$\gamma \circ \iota_* : G_n^o \otimes Z_{n-1}G/G_2^o \to H_2G$$

is the zero map.

Note that we have proved a slightly stronger statement, namely that $\gamma \circ \iota_*^! = 0$, where

$$\iota_*^! : G_n^o \otimes L/G_2^o \to G_n^o \otimes G_{ab} \tag{9.7}$$

and $L \subseteq G$ is the image under g of the centralizer of K_n in K .

<u>COROLLARY 9.4</u>. (Vermani [84]) <u>Let</u> G <u>be nilpotent of class</u> n . <u>Then the sequence</u>

$$G_n^o \otimes G/Z_{n-1}G \overset{\gamma'}{\to} H_2G \to H_2(G/G_n^o) \to G_n^o \to 0 \tag{9.8}$$

<u>is exact</u>.

PROOF: This immediately follows from Proposition 9.3.

PROPOSITION 9.5. (Gaschütz - Neubüser - Yen [31]; Vermani [84]) *Let* G *be nilpotent of class* n . *Then*

$$(9.9)\qquad |H_2G| \leq |G_2^o|^{s(G/ZG)-1}|H_2G_{ab}| ,$$

where $s(Q)$ *denotes the minimal number of generators of* Q .

PROOF: From sequence (9.8) we may infer for $n \geq 2$

$$|H_2G| \leq |H_2(G/G_n^o)|\cdot|G_n^o \otimes (G/Z_{n-1}G)|/|G_n^o| .$$

Now trivially $|G_n^o \otimes (G/Z_{n-1}G)| \leq |G_n^o|^{s(G/Z_{n-1}G)}$, so that

$$|H_2G| \leq |H_2(G/G_n^o)|\cdot|G_n^o|^{s(G/ZG)-1}$$

since $s(G/Z_1G) \geq s(G/Z_{n-1}G)$. Repeating the argument for G/G_n^o , G/G_{n-1}^o , etc. and using

$$s(G/G_k^o/Z(G/G_k^o)) \leq s(G/ZG)$$

we obtain

$$\begin{aligned}|H_2G| &\leq |H_2(G/G_2^o)|\cdot(|G_2^o/G_3^o||G_3^o/G_4^o|\ldots|G_n^o|)^{s(G/ZG)-1}\\ &= |H_2G_{ab}|\cdot|G_2^o|^{s(G/ZG)-1} .\end{aligned}$$

We conclude this section with a proposition on the rank of H_2G for G nilpotent with finite Hirsch number. Recall that the Hirsch number of a nilpotent group G is defined by

$$(9.10)\qquad hG = \sum_{i=1}^{\infty} \operatorname{rank} G_i^o/G_{i+1}^o$$

if all of the (finitely many) summands are finite, and $hG = \infty$ otherwise.

PROPOSITION 9.6. Let G be nilpotent with $hG < \infty$. Then

(9.11) $$\operatorname{rank} H_2G \leq [\operatorname{rank} G_{ab}-1]\cdot hG - \operatorname{rank} H_2G_{ab} .$$

PROOF: Let G be nilpotent of class n. Consider the central extension $E : G_n^o \rightarrowtail G \twoheadrightarrow G/G_n^o$ and the sequence (2.7) associated with E

(9.12) $$G_n^o \otimes G_{ab} \xrightarrow{\gamma} H_2G \to H_2(G/G_n^o) \xrightarrow{\delta_*^E} G_n^o \to 0 ,$$

Counting ranks we obtain

$$\operatorname{rank} H_2G \leq \operatorname{rank} H_2(G/G_n^o) + \operatorname{rank}(G_n^o \otimes G_{ab}) - \operatorname{rank} G_n^o$$
$$\leq \operatorname{rank} H_2(G/G_n^o) + (\operatorname{rank} G_{ab}-1)\cdot \operatorname{rank} G_n^o .$$

Repeating this argument for G/G_n^o, G/G_{n-1}^o, etc. we obtain

$$\operatorname{rank} H_2G \leq \operatorname{rank} H_2G_{ab} + (\operatorname{rank} G_{ab}-1)\cdot \sum_{i=2}^{n} \operatorname{rank} G_i^o/G_{i+1}^o$$
$$= \operatorname{rank} H_2G_{ab}+(\operatorname{rank} G_{ab}-1)hG-(\operatorname{rank} G_{ab}-1)\operatorname{rank} G_{ab}.$$

Finally, since $\operatorname{rank} H_2G_{ab} = \frac{1}{2}(\operatorname{rank} G_{ab}-1)\operatorname{rank} G_{ab}$, we have

$$\operatorname{rank} H_2G \leq (\operatorname{rank} G_{ab}-1)hG - \operatorname{rank} H_2G_{ab}$$

proving our proposition.

We note that the estimate (9.11) is clearly best possible since for G abelian, $hG = \operatorname{rank} G$ and we obtain equality.

V.10. Theorems of Hall's Type

In this section we shall use the methods of this chapter in order to prove Hall's celebrated theorem: Let $H \triangleleft K$. If H and $K/[H,H]$ are nilpotent, then so is K. Also we shall obtain a number of results of the same type. The content of this section is to be found in Robinson [70].

We consider a Σ-group G, i.e. a group G on which the group Σ acts as a group of automorphisms. Let N be a normal Σ-subgroup of G, i.e. a normal subgroup of G which is mapped into itself by every $\sigma \in \Sigma$. We define a series of normal Σ-subgroups of G by setting

$$(10.1) \qquad N_1 = N \ , \ N_{i+1} = [G,N_i] \ , \ i = 1,2,\ldots \ .$$

The quotient of two successive terms N_i/N_{i+1}, $i = 1,2,\ldots$ will be denoted by F_i. Clearly F_i is a Σ-module. Now consider for $i = 2,3,\ldots$ the central extension

$$(10.2) \qquad E_i \ : \ F_{i-1} \rightarrowtail G/N_i \twoheadrightarrow G/N_{i-1} \ .$$

We may thus draw the following diagram (note that for $i \geq 2$, $N_i \subseteq G_2^0$, hence $(G/N_i)_{ab} = G_{ab}$):

$$(10.3) \qquad \begin{array}{ccccccc}
(N_{i-1}/N_{i+1})_{ab} \otimes (G/N_{i+1})_{ab} & & H_2(G/N_{i+1}) & = & H_2(G/N_{i+1}) & & \\
\beta\downarrow & & \downarrow & & \downarrow & & \\
F_{i-1} \otimes G_{ab} & \xrightarrow{\gamma} & H_2(G/N_i) & \rightarrow & H_2(G/N_{i-1}) & \rightarrow & F_{i-1} \rightarrow \cdots \\
 & & \downarrow\delta_* & & \downarrow\delta_* & & \\
 & & F_i & \xrightarrow{0} & F_{i-1} & & \\
 & & \downarrow & & \downarrow & & \\
 & & 0 & & 0 & &
\end{array}$$

Note that N_{i-1}/N_{i+1} lies in the second center of G/N_{i+1} so that we may apply Proposition 2.3 to show that $\delta_*\gamma\beta$ is the commutator map in G/N_{i+1}. In this section we will consider the map

$$(10.4) \qquad \vartheta = \delta_*\gamma \ : \ F_{i-1} \otimes G_{ab} \rightarrow F_i \ .$$

Since by definition $N_i = [G,N_{i-1}] = [N_{i-1},G]$ it follows that ϑ is surjective.

Every term in the diagram (10.3) is a Σ-module, the operation on the tensor product being given via the diagonal (see [43], p.212). It is obvious that every map in the diagram (10.3) is a Σ-module homomorphism. Hence ϑ, as defined in (10.4) is Σ-homomorphic.

PROPOSITION 10.1. *Let* $N \triangleleft G$, $N \subseteq Z_\ell G$ *for some* $\ell \geq 1$. *If* G_{ab} *is a* P-*group for a set* P *of primes, then so is* $[G,N]$.

PROOF: The map $\vartheta : N/[G,N] \otimes G_{ab} \to N_2/N_3$ is epimorphic. Its domain is a P-group, hence so is its range. A finite induction establishes the assertion since $N_i \subseteq Z_{\ell-i+1}$.

PROPOSITION 10.2. (Baer [6]) *Let* $N \triangleleft G$, $N \subseteq Z_\ell G$ *for some* $\ell \geq 1$. *Suppose* N *is torsion free and* G_{ab} *is periodic. Then* $N \subseteq ZG$.

PROOF: The map $\vartheta : N/[G,N] \to N_2/N_3$ is epimorphic. Since its domain is periodic, its range also is. By a finite induction it follows that all factors F_i, $i = 2,3,\ldots$ are periodic. Hence $N_2 = [G,N] \subseteq N$ is periodic; but N is torsion-free so that $[G,N] = e$. Thus $N \subseteq ZG$.

DEFINITION: A class $\mathfrak{C}$ of Σ-modules is called *tensorial* if $A,B \in \mathfrak{C}$ implies that every epimorphic image of $A \otimes B$ is in $\mathfrak{C}$, where Σ acts on the tensor product via the diagonal.

THEOREM 10.3. *Let* $\mathfrak{C}$ *be a tensorial class of modules. If* G *is a* Σ-*group with* $G_{ab} \in \mathfrak{C}$, *then* $G_i^o/G_{i+1}^o \in \mathfrak{C}$ *for* $i = 2,3,\ldots$.

PROOF: Set $N = G$; then $\vartheta : G_{i-1}^o/G_i^o \otimes G_{ab} \to G_i^o/G_{i+1}^o$ is epimorphic. An obvious induction on i establishes the result.

COROLLARY 10.4. *Let* G *be nilpotent. Let* $\mathfrak{B}$ *be a class of groups such that every extension of an* $\mathfrak{B}$-*group by an* $\mathfrak{B}$-*group is an* $\mathfrak{B}$-*group. Suppose that the abelian* $\mathfrak{B}$-*groups form a tensorial class. Then, if*

G_{ab} <u>is in</u> $\mathfrak{B}$, <u>so is</u> G .

<u>PROOF</u>: This immediately follows from Theorem 10.3 , since the lower central series of G is finite and terminates with e .

As specific examples of classes $\mathfrak{B}$ of groups, that satisfy the property required in Corollary 10.4 , we mention the following:

(i) finite P-groups , P a set of primes;

(ii) periodic groups;

(iii) finitely generated groups;

(iv) groups satisfying the maximum condition;

(v) groups satisfying the minimum condition.

Let $\mathfrak{C}$ be a class of Σ-modules. We shall use the following notations.

$P\mathfrak{C}$: The class of Σ-modules having a series of submodules of finite length whose factors belong to $\mathfrak{C}$;

$P'\mathfrak{C}$: The class of Σ-modules having an ascending series of submodules whose factors belong to $\mathfrak{C}$.

An <u>ascending series</u> of A is a set of submodules of A well ordered by inclusion, containing O , A , and all unions of its members.

$Gp\mathfrak{C}$: The class of Σ-groups having a normal series of finite length, whose factors are abelian and belong to $\mathfrak{C}$.

$Gp'\mathfrak{C}$: The class of Σ-groups having an ascending normal series whose factors are abelian and belong to $\mathfrak{C}$.

An ascending normal series of G is a set of subgroups of G well ordered by inclusion, each member normal in the next, containing e , G , and all unions of its members.

LEMMA 10.5. Let $\mathfrak{C}$ be a tensorial class of Σ-modules. Then the class $P\mathfrak{C}$ and $P'\mathfrak{C}$ are tensorial, also.

PROOF: Let (A_i) , $i \in I$, (B_j) , $j \in J$ be (ascending) series of submodules of A , B respectively with A_{i+1}/A_i and B_{j+1}/B_j in $\mathfrak{C}$ for all $i \in I$, $j \in J$. The set $I\times J$ is well-ordered under $(i,j) \leq (i',j')$ if $i \leq i'$ and if $i = i'$, $j \leq j'$. We define an (ascending) series $(A \otimes B)_{(i,j)}$, $(i,j) \in I\times J$ of $A \otimes B$ by

$$(10.5) \qquad (A \otimes B)_{(i,j)} = \mathrm{im}\Big(\sum_{(k,\ell)\leq(i,j)} A_k \otimes B_\ell \to A \otimes B\Big) .$$

Denote by $(i,j)+1$ the first element of the set $\{(m,n) \mid (i,j)<(m,n)\}$. Then it is easy to see that

$$(10.6) \qquad (A \otimes B)_{(i,j)+1}/(A \otimes B)_{(i,j)} = A_{i+1}/A_i \otimes B_{j+1}/B_j .$$

Hence the factors of the (ascending) series $((A \otimes B)_{(i,j)})$, $(i,j) \in I\times J$ of $A \otimes B$ belong to $\mathfrak{C}$. Thus we have shown that $P\mathfrak{C}$, $P'\mathfrak{C}$ are tensorial.

THEOREM 10.6. Let $\mathfrak{C}$ be a tensorial class of Σ-modules which is closed under submodules. If H is a nilpotent normal Σ-subgroup of K such that $K/[H,H]$ is in $Gp\mathfrak{C}$ $(Gp'\mathfrak{C})$, then K is in $Gp\mathfrak{C}$ $(Gp'\mathfrak{C})$, also.

PROOF: Suppose $K/[H,H]$ is in $Gp\mathfrak{C}$. Since $\mathfrak{C}$ is closed under submodules, H_{ab} is in $P\mathfrak{C}$. By Lemma 10.5 and Theorem 10.3 all the factors of the lower central series of H belong to $P\mathfrak{C}$. Hence K belongs to $Gp\mathfrak{C}$. For $Gp'\mathfrak{C}$ the proof is analogous.

We now apply Theorem 10.6 to specific examples for the class $\mathfrak{C}$.

(i) Let $\mathfrak{C}$ be the class of trivial modules $(\Sigma = K)$. Then $Gp\mathfrak{C}$ is the class of nilpotent groups. Thus we obtain

COROLLARY 10.7. (P. Hall [40]) *Let* $H \triangleleft K$, H *nilpotent*, $K/[H,H]$ *nilpotent. Then* K *is nilpotent, also*.

(ii) Let $\mathfrak{C}$ be the class of trivial modules $(\Sigma = K)$. Then $Gp'\mathfrak{C}$ is the class of hypercentral groups. We obtain

COROLLARY 10.8. *Let* $H \triangleleft K$, H *nilpotent*, $K/[H,H]$ *hypercentral. Then* K *is hypercentral*.

(iii) Let $\mathfrak{C}$ be the class of modules $(\Sigma = K)$ whose underlying abelian group is cyclic. Then $Gp\mathfrak{C}$ is the class of supersoluble groups.

COROLLARY 10.9. *Let* $H \triangleleft K$, H *nilpotent*, $K/[H,H]$ *supersoluble. Then* K *is supersoluble*.

(iv) Let $\mathfrak{C}$ be the class of modules $(\Sigma = K)$ whose underlying abelian group is cyclic. Then $Gp'\mathfrak{C}$ is the class of hypercyclic groups.

COROLLARY 10.10. *Let* $H \triangleleft K$, H *nilpotent*, $K/[H,H]$ *hypercyclic. Then* K *is hypercyclic*.

CHAPTER VI

LOCALIZATION OF NILPOTENT GROUPS

In this chapter we present results from the theory of localization and rationalization of nilpotent groups (see Malcev [61], Lazard [52], P. Hall [39], Baumslag [16], Hilton [41]). The approach we have adopted is essentially homological; it owes much to Hilton [41].

Section 1 is preparatory; we show that the integral homology of a P-local abelian group is P-local in positive dimension. Section 2 contains the definition and some basic properties of HPL-groups, i.e. groups whose integral homology is P-local in positive dimension. We show that for an HPL-group the successive quotients of the lower central series are P-local. In Section 3 we consider groups with unique P'-roots. We show that if a group has unique P'-roots, then the successive quotients of the upper central series are P-local. Also, we show that if G is nilpotent, then G has unique P'-roots if and only if G is an HPL-group.

In Section 4 we associate with an arbitrary nilpotent group G an HPL-group G_P. We call G_P the P-localization of G. In Section 5 we derive some of the properties of the P-localization functor. Among others, we prove Malcev's famous theorem that a torsion-free nilpotent group can be embedded in its localization. Also, we show that localization preserves subgroups, normal subgroups, and quotients.

In Section 6 we extend the localization functor to the category of groups that are direct limits of nilpotent groups. Finally we prove in Section 7 a result which has as a corollary the theorem of Baer, which asserts that if G/Z_kG is finite then G^o_{k+1} is also finite.

Much of the material contained in Sections 1 through 5 is to be found in Hilton [41]. Although our presentation differs from [41], many of the proofs have been influenced by or have even been taken from [41]. The material contained in Section 7 owes much to Baumslag [15]. Among other papers on the theory of localization we mention [61], [52], [39], [10], [11], [42], [41], [86].

In this chapter we will be using the theory of spectral sequence, in particular the Lyndon-Hochschild-Serre (L-HS) spectral sequence (see for example [43], Chapter VIII). Also, we shall presume the reader familiar with some basic facts about localization (see for example [2]).

VI.1. Local Abelian Groups

We first recall some basic facts about localization of abelian groups at a family of primes, and then discuss the integral homology of local abelian groups.

Let P denote a (possibly empty) family of primes. By P' we denote the family of primes not in P. The integer m is called a P'-number if m is a product of primes in P'. The ring Z_P is the ring of integers localized at P, i.e. the subring of the rationals consisting of the elements expressible as k/j with j a P'-number. Note that in case P is empty, we have $Z_P = Q$. Recall that for any P the ring Z_P is flat as abelian group.

An abelian group A is called P-local if it is a Z_P-module. Note that if A is P-local, the Z_P-module structure is uniquely determined. If A is any abelian group we may associate with A the P-local group

(1.1) $$A_P = Z_P \otimes A .$$

Clearly $A \longmapsto A_P$ is a functor. The obvious ring homomorphism $Z \longrightarrow Z_P$ induces a map

(1.2) $$\ell : A \to A_P = Z_P \otimes A$$

called the (P-) localization map. It satisfies the following universal property. To any P-local group B and to any homomorphism $f : A \to B$ there exists a unique $f' : A_P \to B$ with $f'\ell = f$. It follows that if A is P-local then $\ell : A \to A_P$ is an isomorphism.

LEMMA 1.1. *Let the diagram*

(1.3) $$\begin{array}{ccccc} A & \rightarrowtail & B & \twoheadrightarrow & C \\ \ell\downarrow & & h\downarrow & & \ell\downarrow \\ A_P & \rightarrowtail & B' & \twoheadrightarrow & C_P \end{array}$$

be commutative with exact rows. Then $B' \cong B_P$ *and* h *is the localization map.*

PROOF: This immediately follows by tensoring (1.3) with Z_P and using the fact that Z_P is flat.

LEMMA 1.2. *Let* A,B *be abelian groups. Then*

(1.4) $$\ell_* : A \otimes B \to A_P \otimes B ,$$

(1.5) $$\ell_* : \mathrm{Tor}(A,B) \to \mathrm{Tor}(A_P,B)$$

are localization maps. If C *is a* P-*local abelian group, then*

(1.6) $$\ell^* : \mathrm{Hom}(A_P,C) \to \mathrm{Hom}(A,C) ,$$

(1.7) $$\ell^* : \mathrm{Ext}(A_P,C) \to \mathrm{Ext}(A,C)$$

are isomorphisms.

PROOF: The first assertion is obvious. To prove (1.5) let $R \rightarrowtail Q \twoheadrightarrow A$ be a free presentation of A. Using (1.4) and the fact that Z_P and hence Q_P is flat we may conclude that the diagram

$$\begin{array}{ccccccccc} 0 \to & Tor(A,B) & \to & R\otimes B & \to & Q\otimes B & \to & A\otimes B & \to 0 \\ & \ell_* \downarrow & & \ell_*^! \downarrow & & \ell_*^! \downarrow & & \ell_*^! \downarrow & \\ 0 \to & Tor(A_P,B) & \to & R_P\otimes B & \to & Q_P\otimes B & \to & A_P\otimes B & \to 0 \end{array}$$

has exact rows and that the maps $\ell_*^!$ are localization maps. Tensoring the upper sequence with Z_P immediately yields the fact that $\ell_* : (Tor(A,B))_P \to Tor(A_P,B)$ is an isomorphism.

The assertion (1.6) follows directly from the universal property of $\ell : A \to A_P$. In order to prove (1.7) we construct an inverse of ℓ^*. Let

(1.7) $$C \rightarrowtail D \twoheadrightarrow A$$

represent an element of $Ext(A,C)$. Then

(1.8) $$Z_P \otimes C \rightarrowtail Z_P \otimes D \twoheadrightarrow Z_P \otimes A$$

is exact. It represents an element in $Ext(A_P,C)$ since $Z_P \otimes C = C$. It is clear that the map that associates with (1.7) the equivalence class of (1.8) is an inverse of $\ell^* : Ext(A_P,C) \to Ext(A,C)$. Thus the proof of Lemma 1.2 is complete.

PROPOSITION 1.3. Let A be an abelian group. Then, for $n \geq 1$, the map $\ell_* : H_nA \to H_n(A_P)$ is the localization map.

PROOF: We first consider the case where A is finitely generated. Let $A = Z/p^kZ$. If $p \notin P$, then $A_P = 0$. If $p \in P$, then $A_P = A$. Thus in both cases $\ell_* : H_nA \to H_n(A_P)$ is the localization map for $n \geq 1$. If $A = Z$, then $H_nA = H_n(A_P) = 0$ for $n \geq 2$, and the assertion is true in this case, also. If the assertion is true for A and B , then it is true for the direct product of A and B by the Künneth exact sequence (II.5.13) and Lemmas 1.1 , 1.2 Thus the proof is complete for A finitely generated. If A is non-finitely generated, it is the direct limit of its finitely generated subgroups. Since both functors $Z_P \otimes -$ and $H_n -$ commute with direct limits the assertion is true in that case, also.

COROLLARY 1.4. *An abelian group* A *is* P-*local if and only if* $H_n A$, $n \geq 1$ *is* P-*local*.

VI.2. Groups with Local Homology

We have seen in Corollary 1.4 that an abelian group A is P-local if and only if $H_n A$ is P-local for $n \geq 1$. It is thus certainly interesting to consider groups G , for which $H_n G$ is P-local for $n \geq 1$. Such a group will be called an HPL-*group*.

PROPOSITION 2.1. *The class of* HPL-*groups is closed under extensions by* HPL-*groups*.

PROOF: Let $E : N \rightarrowtail G \twoheadrightarrow Q$ be an extension of the HPL-group Q by the HPL-group N . We apply the L-HS spectral sequence for the homology of groups to the extension E , and claim that its starting term

$$(2.1) \qquad E_2^{r,s} = H_r(Q, H_s N)$$

is P-local except for $r = s = 0$. To see this we note that the groups $H_s N$ are P-local, except for $s = 0$ since N is an HPL-group. Thus the groups $E_2^{r,s}$ are P-local for $s \geq 1$ since they are computed as homology groups in a complex of P-local abelian groups. Finally, if $s = 0$, then $E_2^{r,s}$ is P-local, except for $r = 0$, since Q is

P-local. It follows that the terms $E_m^{r,s}$, $m \geq 2$ and thus $E_\infty^{r,s}$ are P-local, except for $r = s = 0$. Hence H_nG is P-local, except for $n = 0$.

COROLLARY 2.2. *Let* G *have a series*

$$(2.2) \qquad e = N_{k+1} \subseteq N_k \subseteq \cdots \subseteq N_2 \subseteq N_1 = G$$

of normal subgroups such that the quotients N_i/N_{i+1} , $i = 1,2,\ldots,k$ *are* HPL-*groups. Then* G *is an* HPL-*group*.

PROOF: This follows from Proposition 2.1 by an obvious induction.

PROPOSITION 2.3. *The class of* HPL-*groups is closed under free products and under free products with amalgamated* HPL-*subgroup*.

PROOF: Since the trivial group is an HPL-group it is enough to prove the second assertion. Let U be a HPL-subgroup of the HPL-groups G and K . Then the Mayer-Vietoris sequence (Proposition II.6.1)

$$\cdots \to H_nG \oplus H_nK \to H_n(G *_U K) \to H_{n-1}U \to H_{n-1}G \oplus H_{n-1}K \to \cdots$$

immediately yields that $H_n(G *_U K)$ is P-local for $n \geq 1$.

PROPOSITION 2.4. *Let* G *be an* HPL-*group. Then, for every* $i \geq 1$, G_i^O/G_{i+1}^O *is* P-*local and* G/G_i^O *is an* HPL-*group*.

PROOF: We proceed by induction on i . For $i = 1$ we have $H_1G = G_1^O/G_2^O$. Thus, if G is an HPL-group, G_1^O/G_2^O is P-local. For $i \geq 2$, we consider the extension

$$(2.3) \qquad G_{i-1}^O/G_i^O \rightarrowtail G/G_i^O \twoheadrightarrow G/G_{i-1}^O$$

By Proposition 2.1 we conclude that G/G_i^O is an HPL-group. Consider then the extension $G_i^O \rightarrowtail G \twoheadrightarrow G/G_i^O$ and the associated 5-term sequence

$$(2.4) \qquad H_2G \to H_2(G/G_i^o) \to G_i^o/G_{i+1}^o \to G_{ab} \xrightarrow{\cong} G_{ab} \to 0 .$$

As a cokernel of a homomorphism between P-local abelian groups, the group G_i^o/G_{i+1}^o is P-local, also. The proof of Proposition 2.4 is thus complete.

VI.3. Unique Roots

Let P be a (possibly empty) family of primes. We say that a group G has unique P'-roots if for every $x \in G$ and every P' - number $m \geq 1$ there exists a unique $y \in G$ with $x = y^m$. Note that an abelian group has unique P'-roots if and only if it is P-local.

PROPOSITION 3.1. The class of groups with unique P'-roots is closed under central extensions by P-local abelian groups.

PROOF: Let $N \rightarrowtail G \twoheadrightarrow Q$ be a central extension with Q having unique P'-roots and N being a P-local abelian group. Let $x \in G$ and let $m \geq 1$ be a P'-number. Then there exists $x_1 \in G$ with

$$(3.1) \qquad xN = (x_1N)^m = x_1^m N$$

so that $x = x_1^m \cdot z$ for some $z \in N$. Let $z_1 \in N$ with $z = z_1^m$, then $x = x_1^m \cdot z_1^m = (x \cdot z_1)^m$, since N is central. Thus P'-roots exist. It remains to prove that they are unique. Thus suppose

$$(3.2) \qquad x_1^m = x_2^m \quad , \quad x_1, x_2 \in G .$$

Then $x_1^m N = x_2^m N$, so that by uniqueness of m-th roots in $G/N = Q$ we have $x_1 = x_2\, y$ for some $y \in N$. It follows that

$$(3.3) \qquad x_1^m = (x_2\, y)^m = x_2^m\, y^m$$

whence $y^m = e$. Since N is P-local, it follows that $y = e$. Thus $x_1 = x_2$, so that P'-roots in G are unique.

PROPOSITION 3.2. *Let* G *be a group with unique* P'-*roots. Then* $Z_iG/Z_{i-1}G$, $i \geq 1$ *is* P-*local*.

PROOF: Let $i = 1$. Consider the central extension $ZG \rightarrowtail G \twoheadrightarrow G/ZG$. We shall prove that $ZG = Z_1G/Z_oG$ and G/ZG have unique P'-roots. An obvious induction on i then completes the proof.

Let $x \in ZG$ and let $m \geq 1$ be a P'-number. For all $y \in G$ we have $yxy^{-1} = x$; taking (unique) m-th roots we obtain

$$yx_1y^{-1} = x_1$$

where $x_1^m = x$, so that $x_1 \in ZG$. It follows that ZG has unique m-th roots and hence is P-local. It is clear that G/ZG has m-th roots. To prove uniqueness let

$$x_1^mZG = x_2^mZG \quad , \quad x_1,x_2 \in G . \tag{3.4}$$

Then $x_1^m = x_2^m z$ for some $z \in ZG$. If $z = z_1^m$, we have $x_1 = x_2 z_1$ by uniqueness of m-th roots in G . Since $z_1 \in ZG$ we have

$$x_1ZG = x_2ZG \tag{3.5}$$

so that m-th roots are indeed unique in G/ZG .

The following theorem establishes a close relationship between groups with unique P'-roots and HPL-groups.

THEOREM 3.3. *Let* G *be a nilpotent group. Then the following statements are equivalent*.

(i) G *is an* HPL-*group*;

(ii) G_i^o/G_{i+1}^o , $i \geq 1$ *is* P-*local*;

(iii) G has unique P'-roots;

(iv) $Z_i^o G/Z_{i-1}^o G$, $i \geq 1$ is P-local.

PROOF:

(i) → (ii) follows from Proposition 2.4.

(ii) → (iii). If (ii) holds, the group G is obtained by successive central extensions of a group with unique P'-roots by a P-local abelian group. Thus G has unique P'-roots by Proposition 3.1.

(iii) → (iv) follows from Proposition 3.2.

(iv) → (i). If (iv) holds the group G is obtained by successive (central) extensions of an HPL-group by P-local abelian groups. Thus G is an HPL-group by Proposition 2.1.

We remark that it is not true that the concepts of HPL-groups and groups with unique P'-roots coincide in general. A counterexample is as follows. Let C_P a (multiplicatively written) copy of the additive group of Z_P . Let $t \in C_P$ be a fixed generator of C_P as Z_P-module. Let $G = C_P * C_P$. Then G is a HPL-group by Proposition 2.3. However it is known (see [10]) that there exist elements in G to which there are no P'-roots. Proposition 3.5 will nevertheless establish some kind of freeness property of free products of C_P's with respect to nilpotent HPL-groups. We first prove the following

LEMMA 3.4. Let G be any nilpotent HPL-group and let $x \in G$. Then there is a unique map $f : C_P \to G$ with $f(t) = x$.

PROOF: We first note that uniqueness is clear, since every element in C_P is an m-th root of some power of t where m is a P'-number. To prove existence we proceed by induction on the nilpotency class c of G . If $c = 1$, then G is an abelian HPL-group, hence a P-local abelian group, and the assertion is obvious in this case. Let $c \geq 2$.

Consider the central extension $E : G_c^o \rightarrowtail G \twoheadrightarrow G/G_c^o$. By induction we have a unique map $f' : C_P \to G/G_c^o$ with $f'(t) = xG_c^o$. Consider then a central extension E' with $\Delta[E'] = f'^*\Delta[E] \in H^2(C_P, G_c^o)$ and the diagram

$$\begin{array}{lccccc} E' : & G_c^o & \rightarrowtail & K & \twoheadrightarrow & C_P \\ (3.6) & \| & & g\downarrow & & f'\downarrow \\ E\ : & G_c^o & \rightarrowtail & G & \twoheadrightarrow & G/G_c^o . \end{array}$$

Next we compute $H^2(C_P, G_c^o)$. We first recall that $H^2(C_P, G_c^o) \cong \mathrm{Ext}(Z_P, G_c^o) \oplus \mathrm{Hom}(H_2C_P, G_c^o)$. By Proposition 1.3 we have that $H_2C_P = 0$. By Proposition 2.4 we know that G_c^o is P-local. We may thus apply Lemma 1.2 to prove that $\mathrm{Ext}(Z_P, G_c^o) \cong \mathrm{Ext}(Z, G_c^o) = 0$. It follows that $H^2(C_P, G_c^o) = 0$, so that E' splits, by $s : C_P \to K$, say. We have a map $gs : C_P \to G$ with $gs(t) = x\cdot u$ for some $u \in G_c^o$. Let $f'' : C_P \to G_c^o$ be the map defined by $f''(t) = u$. Then the map $f : C_P \to G$ defined by $f(y) = gs(y)\cdot(f''(y))^{-1}$, $y \in C_P$ is a homomorphism and has the required property.

Let S be a set and let $F = F(S)$ the free group on S . Let $L = L(S) = \underset{s\in S}{\ast}(C_P)_s$, where $(C_P)_s$ is a copy of C_P with distinguished (Z_P-) generator t_s . There is an obvious map $k : F \to L$, defined by $k(s) = t_s$, $s \in S$.

<u>PROPOSITION 3.5</u>. <u>Let</u> G <u>be an</u> HPL-<u>group which is nilpotent of class</u> $\leq c$. <u>Let</u> $f : S \to G$ <u>be a function. Then there are unique homomorphisms</u> $f' : F \to G$, $f'' : L \to G$ <u>such that the diagram</u>

$$(3.7)\qquad \begin{array}{ccc} S & & \\ \cap & \searrow^{f} & \\ F & \overset{f'}{\dashrightarrow} & G \\ k\downarrow & \nearrow_{f''} & \\ L & & \end{array}$$

<u>is commutative. Moreover the diagram</u>

$$(3.8)\qquad \begin{array}{ccccc} F & & \longrightarrow & & F/F^{o}_{c+1} \\ & \searrow^{f'} & & \swarrow^{\bar{f}'} & \\ k\downarrow & & G & & \downarrow k_{c+1} \\ & \nearrow_{f''} & & \nwarrow_{\bar{f}''} & \\ L & & \longrightarrow & & L/L^{o}_{c+1} \end{array}$$

<u>is commutative, also</u>.

<u>PROOF</u>: The statement about f' follows from the universal property of free groups. For every $s \in S$ Lemma 3.4 yields a unique $f_s : (C_P)_s \to G$ with $f_s(t_s) = f(s)$. The universal property of the free product then yields $f'' : L \to G$ making (3.7) commutative. Since $G^{o}_{c+1} = e$ by hypothesis, we obtain $\bar{f}'$, $\bar{f}''$ making (3.8) commutative.

<u>PROPOSITION 3.6</u>. <u>The map</u> $k : F \to L$ <u>has the property that</u> $k_* : H_nF \to H_nL$, $n \geq 1$ <u>is the localization map</u>.

<u>PROOF</u>: For S consisting of one element this follows from Proposition 1.3. For S arbitrary it then is an easy consequence of Corollary II.6.3.

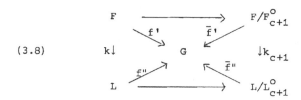

VI.4. Localization of Nilpotent Groups

In this section we shall construct a functor that associates with a nilpotent group G a nilpotent HPL-group G_P and a natural transformation $\ell : G \to G_P$. The construction is in several steps; first we define the group G_P and the homomorphism $\ell : G \to G_P$ (Theorem 4.1). We then define induced homomorphisms which make $-_P$ into a functor and ℓ into a natural transformation (Theorem 4.6).

THEOREM 4.1. *Let* G *be a nilpotent group of class* c. *Then there exists a nilpotent* HPL-*group* G_P *of class* $\leq c$ *and a homomorphism* $\ell : G \to G_P$ *such that*

$$(4.1) \qquad \ell_* : H_n G \to H_n G_P \;, \quad n \geq 1$$

is the localization map.

A homomorphism between not necessarily nilpotent groups $k : G \to K$ with the property that

$$k_* : H_n G \to H_n K \;, \quad n \geq 1$$

is the (P-) localization map is called a (P-) *localization map*. For G nilpotent the map $\ell : G \to G_P$ in Theorem 4.1 is a P-localization map. For short we shall call the specific map constructed in the proof of Theorem 4.1 *the* (P-) localization map (of G).

We shall see in this and the subsequent sections that the localization map $\ell : G \to G_P$ enjoys properties analogous to those of the localization map for abelian groups; in particular it enjoys a universal property with respect to homomorphisms of G into (nilpotent) HPL-groups.

PROOF: (Hilton [41]) We proceed by induction on the nilpotency class c of G. If $c = 1$, then G is abelian. By Proposition 1.3 the localization map $\ell : G \to G_P$ has the required properties.

Let $c \geq 2$. Set $N = G_c^o$, $Q = G/G_c^o$ and consider the central extension $E : N \rightarrowtail G \twoheadrightarrow Q$. Let $\Delta[E] = \xi \in H^2(Q,N)$. Consider the localization maps $\ell' : N \to N_P$ and $\ell'' : Q \to Q_P$ which are given to use by induction. We conclude from the universal coefficient theorem (II.5.1) and Lemmas 1.1, 1.2 that

$$(4.2) \qquad \ell''^* : H^2(Q_P,N_P) \to H^2(Q,N_P)$$

is an isomorphism. We may thus define a central extension

$$(4.3) \qquad E_P : N_P \rightarrowtail G_P \twoheadrightarrow Q_P$$

by $\Delta[E_P] = \zeta = (\ell''^*)^{-1} \circ \ell'_*(\xi) \in H^2(Q_P,N_P)$. Since by construction $\ell''^*(\zeta) = \ell'_*(\xi)$ our Proposition II.4.3 establishes the existence of a map $\ell : G \to G_P$ making the diagram

$$(4.4) \qquad \begin{array}{llll} E & : N \rightarrowtail & G \twoheadrightarrow & Q \\ & \ell' \downarrow & \ell \downarrow & \ell'' \downarrow \\ E_P & : N_P \rightarrowtail & G_P \twoheadrightarrow & Q_P \end{array}$$

commutative. By Proposition 2.1 the group G_P is an HPL-group. In order to prove that $\ell_* : H_nG \to H_nG_P$, $n \geq 1$ is the localization map, we consider the L-HS-spectral sequences of E and E_P. It easily follows from the universal coefficient theorem (II.5.2) and Lemmas 1.1 and 1.2 that

$$(4.5) \qquad H_r(Q,H_sN) \xrightarrow{\ell'_*} H_r(Q,H_sN_P) \xrightarrow{\ell''_*} H_r(Q_P,H_sN_P)$$

is the localization map, except for $r = s = 0$. It follows that the E_2-term and hence the E_∞-term is localized by the map induced by $\ell''_*\ell'_*$, except for $r = s = 0$. Hence $\ell_* : H_n G \to H_n G_P$ is the localization map, except for $n = 0$.

It is clear from the construction of G_P that if G is nilpotent of class c then G_P is also nilpotent of class $\leq c$. The proof of Theorem 4.1 is thus complete.

We remark that it follows from Proposition 2.4 that if G is a nilpotent HPL-group of class $\leq c$ then G_c^o is P-local and G/G_c^o is an HPL-group. Thus we may conclude that if G is a nilpotent HPL-group then $\ell : G \to G_P$ is an isomorphism. Of course, in this case we will use the map ℓ to identify the groups G and G_P .

Note that no special properties of the lower central series are used in the proof of Theorem 4.1. In order to obtain, for a nilpotent group G a localization map, i.e. an HPL-group K and a map $k : G \to K$ such that $k_* : H_n G \to H_n K$, $n \geq 1$ is the localization map we may use <u>any</u> finite central series whatsoever. We shall see later (Corollary 5.3) that the construction is indeed independent of the chosen central series. However, for certain results to be proved later, it is convenient to use the <u>lower</u> central series for the construction.

<u>PROPOSITION 4.2</u>. <u>Let</u> G <u>be a nilpotent group and let</u> $\ell : G \to G_P$ <u>be the localization map. If</u> $x \in G_P$ <u>there exists a</u> P'-<u>number</u> $m \geq 1$ <u>with</u> $x^m \in \ell G$.

<u>PROOF</u>: We proceed by induction on the nilpotency class c of G . If $c = 1$, then G is abelian, and our assertion is well-known in that case. Let $c \geq 2$. Consider the diagram

$$(4.7)\qquad \begin{array}{lcccccc} E & : & N & \rightarrowtail & G & \twoheadrightarrow & Q \\ & & \ell'\downarrow & & \ell\downarrow & & \ell''\downarrow \\ E_P & : & N_P & \rightarrowtail & G_P & \twoheadrightarrow & Q_P \end{array}$$

where we have used the notation introduced in the proof of Theorem 4.1. Let $x \in G_P$, then there exists a P'-number $m_1 \geq 1$ and $y \in G$ with

$$x^{m_1}N_P = (xN_P)^{m_1} = \ell''(yN) = (\ell y)N_P .$$

Thus there exists $z \in N_P$ with $x^{m_1} = (\ell y)z$. But then we may find a P'-number m_2 and $u \in N$ with $z^{m_2} = \ell'(u)$. Thus

$$(4.8)\qquad x^{m_1 \cdot m_2} = (\ell y)^{m_2} \cdot z^{m_2} = \ell(y^{m_2}) \cdot \ell'(u) = \ell(y^{m_2} \cdot u) ,$$

so that $m = m_1 \cdot m_2$ has the required properties.

PROPOSITION 4.3. (Hilton [41]) *Let G be a nilpotent group and let $\ell : G \to G_P$ be the localization map. Then $x \in \ker \ell$ if and only if x is a P'-torsion element.*

PROOF: Let $x \in G$ be a P'-torsion element. Then $y = \ell(x)$ is a P'-torsion element, also. Thus there exists a P'-number $m \geq 1$ with $y^m = e$. Uniqueness of m-th roots in G_P then yields $y = e$.

To prove the converse we proceed by induction on the nilpotency class c of G. If $c = 1$, then G is abelian and the result is well-known in that case. Let $c \geq 2$, and let $x \in \ker \ell$. Using the notation introduced in the proof of Theorem 4.1 and refering to the diagram (4.7) we have $xN \in \ker \ell''$. Thus by induction xN is a P'-torsion element in Q, so that there exists a P'-number $m \geq 1$ with $x^m \in N$. Since $x \in \ker \ell$, it follows that $x^m \in \ker \ell'$. Hence $x^m \in N$ is a P'-torsion element, so that x itself is a P'-torsion element.

COROLLARY 4.4. *Let* G *be a torsion-free nilpotent group. Then* $\ell : G \to G_P$ *is injective for any* P .

PROOF: This immediately follows from Proposition 4.3.

We note that for $P = \emptyset$, Corollary 4.4 is a famous result of Malcev [61].

PROPOSITION 4.5. *Let* $\ell : G \to K$ *be a localization map. Then the maps*

$$k_i : G_i^o/G_{i+1}^o \to K_i^o/K_{i+1}^o \ , \tag{4.9}$$

$$\ell_i : G/G_i^o \to K/K_i^o \tag{4.10}$$

induced by ℓ *are the localization maps for all* i .

Note that we do not suppose G (or K) nilpotent.

PROOF: We proceed by induction on i . For $i = 1$ the assertion follows from the fact that

$$\ell_* : H_1G \to H_1K$$

is the localization map by hypothesis. Let $i \geq 2$. Consider the diagram

$$\begin{array}{ccccc} G_{i-1}^o/G_i^o & \rightarrowtail & G/G_i^o & \twoheadrightarrow & G/G_{i-1}^o \\ k_{i-1}\downarrow & & \ell_i\downarrow & & \ell_{i-1}\downarrow \\ K_{i-1}^o/K_i^o & \rightarrowtail & K/K_i^o & \twoheadrightarrow & K/K_{i-1}^o \end{array} \tag{4.11}$$

where k_{i-1} and ℓ_{i-1} are localization maps by induction. By construction of the localization of G/G_i^o it follows that ℓ_i is the localization map. Consider then the diagram

$$(4.12)\qquad \begin{array}{ccccc} G_i^o & \rightarrowtail & G & \twoheadrightarrow & G/G_i^o \\ \downarrow & & k\downarrow & & \ell_i\downarrow \\ K_i^o & \rightarrowtail & K & \twoheadrightarrow & K/K_i^o \end{array}$$

and the associated 5-term sequences

$$(4.13)\qquad \begin{array}{ccccccc} H_2G & \to & H_2(G/G_i^o) & \to & G_i^o/G_{i+1}^o & \to & 0 \\ \ell_*\downarrow & & (\ell_i)_*\downarrow & & k_i\downarrow & & \\ H_2K & \to & H_2(K/K_i^o) & \to & K_i^o/K_{i+1}^o & \to & 0 \end{array}$$

Since ℓ_* and $(\ell_i)_*$ are localization maps, it follows that k_i is the localization map, thus completing the proof.

THEOREM 4.6. *Let* $g : G \to K$ *be a homomorphism of nilpotent groups. There is precisely one homomorphism* $g_P : G_P \to K_P$ *such that the following square is commutative*

$$(4.14)\qquad \begin{array}{ccc} G & \xrightarrow{g} & K \\ \ell\downarrow & & \ell\downarrow \\ G_P & \xrightarrow{g_P} & K_P \end{array}$$

PROOF: We first note that if it exists g_P must be unique. For it follows from Proposition 4.2 that if $x \in G_P$ there exists a P'-number $m \geqslant 1$ and $y \in G$ with $x^m = \ell y$. Thus $g_P(x^m) = \ell g(y)$, so that $g_P(x)$ must be the unique m-th root of $\ell g(y)$ in K_P.

It remains to prove that g_P exists. Choose a free presentation $f : F \twoheadrightarrow G$ of G. Then Proposition 3.5 yields a commutative diagram

$$(4.15)\qquad \begin{array}{ccccccc} F & \rightarrow & F/F_{c+1} & \xrightarrow{f'} & G & \xrightarrow{g} & K \\ k\downarrow & & k_{c+1}\downarrow & & \ell\downarrow & & \ell\downarrow \\ L & \rightarrow & L/L_{c+1} & \xrightarrow{\bar f''} & G_P & & \\ & & & \xrightarrow[\bar f''_1]{} & & & K_P \end{array}$$

It follows from Proposition 4.5 that k_{c+1} is the localization map. Also, it follows from Proposition 4.2 that $\bar f''$ is surjective. In order to define $g_P : G_P \to K_P$ it is thus enough to prove that $\bar f''_1(x) = e$ if $x \in \ker \bar f''$. To prove this we use Proposition 4.2 to find a P'-number m with $x^m = k_{c+1}(y)$ with $y \in F/F_{c+1}$. Then $\ell f'(y) = \bar f''(x^m) = (\bar f''(x))^m = e$. We conclude from Proposition 4.3 that $f'(y)$ is a P'-torsion element. Thus $gf'(y)$ is a P'-torsion element, so that $e = \ell g f'(y) = \bar f''_1(x)$. The proof that $\ker \bar f'' \subseteq \ker \bar f''_1$ is thus complete. We define $f_P : G_P \to K_P$ to be the map induced by $\bar f''_1$. It is then clear that (4.14) is commutative.

VI.5. Properties of Localization of Nilpotent Groups

We shall first state a universal property of the map ℓ as defined in Section 4, and then draw a number of consequences.

THEOREM 5.1. *Let* G *be a nilpotent group. The localization map* $\ell : G \to G_P$ *has the following universal property. To any nilpotent* HPL-*group* K *and to any homomorphism* $f : G \to K$ *there exists a unique* $f' : G_P \to K$ *with* $f'\ell = f$.

PROOF: Since $K = K_P$ we may, and indeed must define $f' = f_P : G_P \to K_P$.

COROLLARY 5.2. *Let* G *and* K *be nilpotent and let* $k : G \to K$ *be a localization map. Then there exists a uniquely determined isomorphism* $k' : G_P \xrightarrow{\sim} K$ *with* $k'\ell = k$.

PROOF: Since K is an HPL-group, Theorem 5.1 yields a map $k' : G_P \to K$ with $k'\ell = k$. Since $k_* : H_1G \to H_1K$ and $k_* : H_2G \to H_2K$ are the localization maps, we have

$$k'_* : H_1G_P \cong H_1K \quad , \quad k'_* : H_2G_P \cong H_2K \; .$$

It then follows from Corollary IV.1.2 that k' is an isomorphism.

With this result we are able to substantiate the remark after the proof of Theorem 4.1. Let G be a nilpotent group. Suppose we are given a finite central series of G. Using this central series instead of the lower central series in the construction described in the proof of Theorem 4.1 we obtain an HPL-group K, say and a localization map $k : G \to K$. It then follows from Corollary 5.2 that there is a uniquely determined isomorphism $k' : G_P \cong K$ with $k'\ell = k$. We may thus state

COROLLARY 5.3. *Let* G *be nilpotent. Let* $k : G \to K$ *be a localization map constructed using any finite central series of* G. *Then there exists a uniquely determined isomorphism* $k' : G_P \cong K$ *with* $k'\ell = k$.

COROLLARY 5.4. *Let* G *and* K *be nilpotent, and let* $h : G \to K$ *be a homomorphism. Suppose that* K *has unique* P'-*roots, that* $\ker h$ *is the* P'-*torsion subgroup of* G, *and that to every* $x \in K$ *there exists a* P'-*number* $m \geq 1$ *with* $x^m \in hG$. *Then there exists a unique isomorphism* $h' : G_P \cong K$ *with* $h'\ell = h$.

PROOF: Since K has unique P'-roots it is an HPL-group by Theorem 3.3. By Theorem 5.1 there exists a unique homomorphism $h' : G_P \to K$ with $h'\ell = h$. It remains to prove that h' is an isomorphism. Thus let $y \in \ker h'$. By Proposition 4.2 there exists a P'-number $m \geq 1$ with $y^m \in \ell G$. Let $y^m = \ell z$, $z \in G$; then

$$hz = h'\ell z = h'(y^m) = (h'y)^m = e \ ,$$

so that z is a P'-torsion element. By Proposition 4.3 we have $\ell z = e$. But then $y^m = e$ and uniqueness of P'-roots yields $y = e$. Thus h' is injective. To prove surjectivity let $u \in K$. Then there exists a P'-number with $u^m = hv$ for some $v \in G$. Consider $w \in G_P$ with $w^m = \ell v$. Then clearly $h'w = u$ by uniqueness of P'-roots, thus completing the proof.

We remark that Corollary 5.4 constitutes the basis of Hilton's approach [41].

PROPOSITION 5.5. *Let* G *be nilpotent, and let* $g : G \to Q$ *be an epimorphism. Then* $g_P : G_P \to Q_P$ *is surjective*.

PROOF: Let $x \in Q_P$. By Proposition 4.2 there exists a P'-number $m \geqslant 1$ with $x^m = \ell y$ for some $y \in Q$. Let $gv = y$, $v \in G$, and define $u \in G_P$ by $u^m = \ell v$. Then clearly $g_P(u) = x$. Thus g_P is surjective.

PROPOSITION 5.6. *Let* G *be nilpotent and let* $h : N \to G$ *be a monomorphism. Then* $h_P : N_P \to G_P$ *is injective*.

PROOF: Suppose $x \in \ker h_P$. Then there exists a P'-number $m \geqslant 1$ with $x^m = \ell y$ for some $y \in N$. Thus

$$\ell hy = h_P \ell y = h_P(x^m) = (h_P x)^m = e \ ,$$

so that hy must be a P'-torsion element by Proposition 4.3. Since h is monomorphic, y is a P'-torsion element. Hence $y \in \ker \ell$, so that $x^m = e$. By uniqueness of m-th roots in N_P we have $x = e$.

COROLLARY 5.7. *Let* $N \overset{h}{\rightarrowtail} G \overset{g}{\twoheadrightarrow} Q$ *be an extension with* G *nilpotent. Then the sequence*

$$N_P \overset{h_P}{\rightarrowtail} G_P \overset{g_P}{\twoheadrightarrow} Q_P$$

is an extension.

PROOF: By Propositions 5.5 and 5.6 we know that h_P is injective and that g_P is surjective. Next we show that the composition $g_P h_P$ is the trivial map. Thus let $x \in N_P$. Then there exists a P'-number $m \geq 1$ with $x^m = \ell y$ for some $y \in N$. Thus

$$e = \ell g h y = g_P h_P \ell y = g_P h_P x^m = (g_P h_P x)^m$$

whence it follows that $g_P h_P x = e$. In order to complete the proof it remains to show that the kernel of g_P is contained in N_P. Thus let $u \in \ker g_P$. Then there exists a P'-number m with $u^m = \ell v$ for some $v \in G$. Since

$$\ell g v = g_P \ell v = g_P(u^m) = (g_P u)^m = e ,$$

the element $gv \in Q$ is a P'-torsion element. Thus there exists a P'-number m' with $(gv)^{m'} = g(v^{m'}) = e$. It follows that there exists $y \in N$ with $hy = v^{m'}$. Define $x \in N_P$ by $x^{m \cdot m'} = \ell y$. Then

$$(h_P x)^{m \cdot m'} = h_P(x^{m \cdot m'}) = h_P \ell y = \ell h y = \ell(v^{m'}) = u^{m \cdot m'}$$

so that $h_P x = u$, as required.

PROPOSITION 5.8. Let G be a finitely generated nilpotent group and let U be a subgroup of G. Then $U_P \xrightarrow{\cong} G_P$ if and only if $[G:U] = m$ is a P'-number.

PROOF: Suppose $[G:U] = m$ is a P'-number. Since G is nilpotent, the normalizer of U in G is bigger than U, so that it suffices to consider the case where U is normal in G. Thus let

$$U \rightarrowtail G \twoheadrightarrow Q$$

be an extension with Q finite of order m where m is a P'-number. By Corollary 5.7 localizing yields an isomorphism $U_P \xrightarrow{\sim} G_P$. (Note that for this part of the assertion we did not need the fact that G is finitely generated.)

Conversely, let $U_P \xrightarrow{\sim} G_P$ be an isomorphism. Again we may suppose U normal in G . Consider the diagram

$$\begin{array}{ccccc} U & \rightarrowtail & G & \twoheadrightarrow & Q \\ \downarrow & & \downarrow & & \downarrow \\ U_P & \overset{\sim}{\rightarrowtail} & G_P & \twoheadrightarrow & Q_P \end{array}$$

By Corollary 5.7 we have $Q_P = e$ whence it follows that Q consists of P'-torsion elements. Since G is finitely generated, Q is. Hence Q is finite of an order m where m is a P'-number.

<u>PROPOSITION 5.9</u>. <u>Let</u> $E : N \rightarrowtail G \twoheadrightarrow Q$ <u>be an extension with</u> N <u>a nilpotent</u> HPL-<u>group and</u> Q <u>a finite group of order</u> m <u>where</u> m <u>is a</u> P'-<u>number. Then</u> E <u>splits</u>.

<u>PROOF</u>: We proceed by induction on the nilpotency class c of N . If N is abelian, then $H^2(Q,N) = 0$. For $H^2(Q,N)$ is P-local, since N is P-local, and it is of exponent m , since Q is of order m . Thus $H^2(Q,N) = 0$. Let $c \geq 2$. Consider the diagram

$$\begin{array}{ccccc} N & \rightarrowtail & G & \twoheadrightarrow & Q \\ \downarrow & & g\downarrow & & \| \\ N/N_c & \rightarrowtail & G/N_c & \twoheadrightarrow & Q \end{array}$$

By induction the lower sequence splits, by $s_1 : Q \to G/N_c$ say. Consider s_1Q and $g^{-1}(s_1Q)$. This yields the extension

$$E' : N_c \rightarrowtail g^{-1}(s_1Q) \twoheadrightarrow s_1Q \ .$$

Since N_c is P-local and s_1Q is of order m, the extension E' splits by $s_2 : s_1Q \to g^{-1}(s_1Q)$, say. Composition yields a splitting s of E

$$s : Q \xrightarrow{s_1} s_1Q \xrightarrow{s_2} g^{-1}(s_1Q) \subseteq G \ ,$$

thus completing the proof.

COROLLARY 5.10. *Let* $E : N \rightarrowtail G \twoheadrightarrow Q$ *be an extension with* N *a nilpotent* HPL-*group and* Q *a finite group of order* m *where* m *is a* P'-*number. Suppose* $N \subseteq Z_kG$. *Then* $G = N \times Q$.

PROOF: By Proposition 5.9 we know that E splits. Let $s : Q \to G$ be a splitting. The conjugation in G induces an operation of Q on N. It is to prove that this operation is trivial. Suppose not. Since Q operates trivially on $Z_1G \cap N$ there exists $1 < i < k$ such that Q operates trivially on $Z_iG \cap N$ but non-trivially on $Z_{i+1}G \cap N$. It follows that there exists $x \in Z_{i+1}G \cap N$ with $[x,sQ] \neq e$. Since $[Z_{i+1}G,G] \subseteq Z_iG$ it is easy to see that the function $q : Q \to N$ defined by $q(y) = [x,sy]$, $y \in Q$ is a homomorphism. Since q is non-trivial, the image of Q is a non-trivial finite subgroup of N of order m', say, where m' is a P'-number. Since N is a nilpotent HPL-group this is a contradiction.

PROPOSITION 5.11. *Let* U *be a subgroup of the nilpotent group* G . *Then the set* J *of all elements* $x \in G$ *such that there is a* P'-*number* $m \geq 1$ *with* $x^m \in U$ *is a subgroup of* G .

PROOF: Using the P-localization map ℓ the set J may be described by

$$J = \ell^{-1}(U_P \cap \ell G) .$$

It is then clear that J is a subgroup.

The group J is usually called the P'-isolator of U in G (see [15]).

VI.6. Localization of Non-Nilpotent Groups

In this section we shall say a few words on two classes of (not necessarily nilpotent) groups whose members admit localization maps in a natural way.

We start with the following remark. Let G be a group. We are looking for an HPL-group K and a homomorphism $\ell : G \to K$ which is a localization map, i.e. a map with $\ell_* : H_n G \to H_n K$, $n \geq 1$ the localization map. It is clear that in general K is not uniquely determined. However, it follows from Proposition 4.5 that the homomorphisms

$$(6.1) \qquad \ell_i : G/G_i^o \to K/K_i^o$$

are localization maps for all i. Hence at least the quotients of K by the terms of the lower central series are determined. For the classes of groups we intend to study in this section it is possible to define a localization map in a natural way.

We first consider the class of groups G which are direct limits of nilpotent groups G^i, i.e.

$$(6.2) \qquad G = \varinjlim G^i .$$

It is obvious that instead of G^i we may consider its canonical (nilpotent) image in G. Hence we may suppose without loss of generality that all G^i are subgroups of G, and that G is the direct limit of (all of) its nilpotent subgroups. We define the localization G_P of G by

(6.3) $$G_P = \varinjlim (G^i)_P .$$

The maps $\ell^i : G^i \to (G^i)_P$ yield a homomorphism $\ell : G \to G_P$. Since both functors H_n - as well as $\mathbb{Z}_P \otimes$ - commute with direct limits, we immediately have

PROPOSITION 6.1. The map $\ell_* : H_nG \to H_nG_P$, $n \geq 1$ is the localization map, so that $\ell : G \to G_P$ is a localization map.

Let G and K be two groups in our class and let $f : G \to K$ be a homomorphism. Then f induces a map of the directed systems $\{G^i\}$, $\{K^i\}$ and thus a well-defined map of the directed systems $\{(G^i)_P\}$, $\{(K^i)_P\}$. Hence, we obtain by universality a map $f_P : G_P \to K_P$. We have

PROPOSITION 6.2. The localization $G \mapsto G_P$ is a functor and $\ell : G \to G_P$ is a natural transformation.

PROOF: It is obvious that f_P as defined above satisfies the required commutativity relations.

PROPOSITION 6.3. Let $\ell : G \to G_P$ the localization. Then G_P has unique P'-roots. Moreover, we have that to $x \in G_P$ there exists a P'-number $m \geq 1$ with $x^m \in \ell G$. Finally, $y \in \ker \ell$ if and only if y is a P'-torsion element.

PROOF: Let $G^i \subseteq G^j$ be two nilpotent subgroups of G . By Proposition 5.6 the induced map $(G^i)_P \to (G^j)_P$ is injective. It follows that

(6.4) $$(G^i)_P \subseteq G_P .$$

Now let $x \in G_P = \varinjlim (G^i)_P$; then there exists $i \in I$ with $x \in (G^i)_P$ and $(G^i)_P$ has unique P'-roots. Also there exists a P'-number $m \geq 1$ with $x^m \in \ell G^i$. Thus $x^m \in \ell G$. Finally, let $y \in G$ be a P'-torsion element. Then there exists $i \in I$ with $y \in G^i$. Thus $y \in \ker(\ell^i : G^i \to (G^i)_P)$. It follows that $y \in \ker \ell$. Conversely, if $y \in \ker \ell$, then $y \in \ker \ell^i$ for some $i \in I$. Hence y is a P'-torsion element.

We now turn to free products with amalgamated subgroups.

PROPOSITION 6.4. *Let* G , K *be two groups, that are unions of their nilpotent subgroups. Let* U *be a subgroup of both* G *and* K . *Then there is a map* $\ell : G *_U K \to G_P *_{U_P} K_P$ *and* ℓ *is a localization map.*

PROOF: First we note that U is also the union of its nilpotent subgroups. Since the localization maps $\ell : G \to G_P$ and $\ell : K \to K_P$ have just the P'-torsion elements as kernels it follows that $\ker(\ell : U \to U_P) = \ker(\ell : G \to G_P) \cap U = \ker(\ell : K \to K_P) \cap U$. It follows that U_P is a subgroup of both G_P and K_P . The existence of $\ell : G *_U K \to G_P *_{U_P} K_P$ is then obvious from universality. Finally, there is a commutative diagram with exact rows by Proposition II.6.1

$$
(6.5)\qquad
\begin{array}{ccccccc}
\cdots \to & H_n G \oplus H_n K & \to & H_n(G *_U K) & \to & H_{n-1} U \to \cdots \\
& \alpha\downarrow & & \beta\downarrow & & \gamma\downarrow \\
\cdots \to & H_n G_P \oplus H_n K_P & \to & H_n(G_P *_{U_P} K_P) & \to & H_{n-1} U_P \to \cdots
\end{array}
$$

Since α , γ are localization maps, β is.

For free products Proposition 6.5 may be sharpened.

PROPOSITION 6.5. *Let* $\ell : G \to G_P$ *and* $\ell : K \to K_P$ *be localization maps. Then the obvious map* $\ell : G * K \to G_P * K_P$ *is a localization map, also.*

PROOF: This is obvious since

$$(6.6) \qquad \ell_* : H_n(G*K) = H_nG \oplus H_nK \to H_nG_P \oplus H_nK_P = H_n(G_P*K_P)$$

is the localization map for $n \geq 1$.

COROLLARY 6.6. *Let* G , K *be in* $\underline{\underline{V}} = \underline{\underline{N}}_c$. *Then the obvious map* $\ell : G *_V K \to G_P *_V K_P$ *is the localization map*.

PROOF: By definition $G *_V K = G*K/(G*K)^{o}_{c+1}$. By Propositions 6.5 and 4.5 the assertion is then obvious.

VI.7. A Result on Extensions of Homomorphisms

In this section we shall prove a result on extensions of homomorphisms that has as an immediate corollary a well-known theorem of Baer [7].

We start with a number of remarks. Let N be a normal subgroup of G . Suppose $N \subseteq Z_kG$ for some k , so that in particular N is nilpotent. For $x \in G$ let $x : N \to N$ denote the induced automorphism. For any set P of primes the maps $x_P : N_P \to N_P$ with

$$(7.1) \qquad \begin{array}{ccc} N & \xrightarrow{x} & N \\ \ell\downarrow & & \ell\downarrow \\ N_P & \xrightarrow{x_P} & N_P \end{array}$$

commutative define an action of G on N_P . The localization map ℓ is then compatible with this G-action.

LEMMA 7.1. *Let* M *be a nilpotent* HPL-*group on which* G *acts as group of automorphisms. Let* $f : N \to M$ *be a homomorphism compatible with the* G-*action on* N *and* M . *Then the map* $f' : N_P \to M$ *with*

$$\begin{array}{ccc} N & \xrightarrow{\ \ell\ } & N_P \\ f\downarrow & \swarrow_{f'} & \\ M & & \end{array}$$

commutative is also compatible with the G-*action*.

PROOF: Let $u \in N_P$ then there exists a P'-number $m \geq 1$ with $u^m = \ell v$ for some $v \in N$. For $x \in G$ we have

$$(x \circ f'u)^m = x \circ f'(u^m) = x \circ f'\ell v = x \circ fv = f(x \circ v) =$$
$$= f'\ell(x \circ v) = f'(x \circ \ell v) = f'(x \circ u^m) = (f'(x \circ u))^m .$$

Taking (unique) m-th roots establishes the result.

THEOREM 7.2. *Let* N *be a normal subgroup of* G *with* $N \subseteq Z_k G$ *for some* k . *Suppose that* $Q = G/N$ *is finite of order* m *where* m *is a* P'-*number. Let* M *be a nilpotent* HPL-*group on which* G *acts and let* $f : N \to M$ *be a homomorphism compatible with the* G-*action on* N *and* M . *Then there exists a unique homomorphism* $f' : G \to M$ *with*

$$(7.2) \qquad \begin{array}{ccc} N & \overset{h}{\rightarrowtail} & G \\ f\downarrow & \swarrow_{f'} & \\ M & & \end{array}$$

commutative.

PROOF: First we remark that by Lemma 7.1 it is enough to prove the result for $M = N_P$ and $f : N \to M$ the localization map. Next we note that uniqueness is clear, since for $x \in G$ we have $x^m \in N$, so that $f(x)$ is the unique m-th root of $f(x^m)$ in M .

To prove the existence we consider the series

$$(7.3) \qquad N_1 = N , \; N_{i+1} = [G,N_i] \quad , \; i = 1,2,\ldots .$$

Since $N_\ell \subseteq Z_{k-\ell+1}G$, $\ell \geq 1$ it follows that $N_{k+1} = e$. Thus the series $\{N_i\}$ is a finite central series of N . By Corollary 5.3 it may be used to construct the localization map $\ell : N \to N_P$. Using Proposition 5.6 we may define subgroups of $M = N_P$ by $M_i = (N_i)_P$. Note that by construction every $x \in G$ operates trivially on N_i/N_{i+1} and hence on M_i/M_{i+1} . We now proceed by induction on the length c of the series $\{N_i\}$. If $c = 1$, then N is central. We may consider

$$(7.4) \qquad \begin{array}{ccccccccc} H_2G & \to & H_2Q & \xrightarrow{\delta} & N & \xrightarrow{h_*} & G_{ab} & \to & Q_{ab} \to 0 \\ & & & & \ell\downarrow & \swarrow_{f''} & & & \\ & & & & N_P & & & & \end{array}$$

Since Q is of order m , the group H_2Q is P'-torsion, so that $\ell\delta = 0$. Thus the localization map $\ell : N \to N_P$ factors through coker $\delta = \text{im}\, h_* \subseteq G_{ab}$. Since Q_{ab} is of order dividing m there exists $f'' : G_{ab} \to N_P$ and hence $f' : G \to N_P$ such that (7.2) is commutative. This establishes the assertion in case $c = 1$.

Now let $c \geq 2$. Consider the diagram

$$(7.5) \qquad \begin{array}{lccccc} F : & N_c & \rightarrowtail & N & \twoheadrightarrow & N/N_c \\ & \Vert \searrow \ell' & & h\downarrow \searrow \ell & & h''\downarrow \searrow \ell'' \\ E : & N_c & \rightarrowtail & G & \twoheadrightarrow & G/N_c \\ & \ell' \searrow & & f \searrow & & f'' \searrow \\ F_P : & M_c & \rightarrowtail & M & \twoheadrightarrow & M/M_c \end{array}$$

where ℓ , ℓ' , ℓ'' denote various localization maps. Suppose f'' is given by induction such that the right most triangle is commutative, i.e. that $f''h'' = \ell''$. We have to find $f : G \to M$ such that (7.5) is commutative, in particular that $fh = \ell$. The construction of f is in two steps; we shall first assert the existence of $\bar{f} : G \to M$ with the property that the bottom of (7.5) commutes. Secondly we shall use

$\bar{f}$ to construct $f : G \to M$ such that the whole of (7.5) commutes.

The maps of (7.5) which are already known yield the commutative diagram

(7.6)
$$\begin{array}{ccccc} H^2(G/N_c,N_c) & \xrightarrow{\ell^!_*} & H^2(G/N_c,M_c) & \xleftarrow{f''*} & H^2(M/M_c,M_c) \\ h''*\downarrow & & \downarrow h''* & & \| \\ H^2(N/N_c,N_c) & \xrightarrow{\ell^!_*} & H^2(N/N_c,M_c) & \xleftarrow{\ell''*} & H^2(M/M_c,M_c) \end{array}$$

In order to assert the existence of a map $\bar{f} : G \to M$ such that the bottom of (7.5) commutes we have to verify that

(7.7) $$\ell^!_*(\Delta[E]) = f''*(\Delta[F_P]) .$$

From (7.6) we know that

$$\begin{aligned} h''*\ell^!_*(\Delta[E]) &= \ell^!_* h''*(\Delta[E]) \\ &= \ell^!_*(\Delta[F]) \\ &= \ell''*(\Delta[F_P]) \\ &= h''*f''*(\Delta[F_P]) \end{aligned}$$

To prove (7.7) it is thus certainly enough to show that $h''*$ is monomorphic. But since N/N_c is of index m in G/N_c, we have that

$$\mathrm{Cor} \circ \mathrm{Res} : H^2(G/N_c,M_c) \to H^2(G/N_c,M_c)$$

is multiplication by m. Since M_c and hence $H^2(G/N_c,M_c)$ is P-local, m is invertible in $H^2(G/N_c,M_c)$ and $\mathrm{Cor} \circ \mathrm{Res}$ is an isomorphism. It follows that $\mathrm{Res} = h''*$ is monomorphic, thus completing the proof of (7.7) and establishing the existence of $\bar{f} : G \to M$.

It remains to show that there exists $f : G \to M$ such that (7.5) is commutative. Consider

$$
(7.8)\qquad
\begin{array}{ccccc}
N_c & \rightarrowtail & N & \twoheadrightarrow & N/N_c \\
\| & & h\downarrow & & h''\downarrow \\
N_c & \rightarrowtail & G & \twoheadrightarrow & G/N_c \\
\ell'\downarrow & & \bar{f}\downarrow & & f''\downarrow \\
M_c & \rightarrowtail & M & \twoheadrightarrow & M/M_c \\
\| & & \downarrow q\cdot 1_M & & \| \\
M_c & \rightarrowtail & M & \twoheadrightarrow & M/M_c
\end{array}
$$

Recall first that $f''h'' = \ell'' : N/N_c \to M/M_c$. Let $x \in N$. Then $\bar{f}h(x)\cdot M_c = \ell''(xN_c) = \ell(x)\cdot M_c$ where $\ell : N \to M$ is the localization map. Thus there exists $y \in M_c$ with

$$(7.9)\qquad (\bar{f}h(x))\cdot y = \ell(x) .$$

Define $\bar{q} : N \to M$ by $\bar{q}(x) = y \in M$. Since $y \in M_c$ and M_c is central, $\bar{q}$ is homomorphic. Since M is P-local, $\bar{q}$ factors as $\bar{q} = q\circ\ell$ where $q : M \to M$. Clearly $\bar{q}$ vanishes on N_c , whence it follows that q vanishes on M_c . Define $f : G \to M$ by

$$f(z) = \bar{f}(z)\cdot q(\bar{f}(z)) , \quad z \in G$$

It is easy to see that f is homomorphic. It remains to show that (7.5) is commutative, i.e. that $fh = \ell$. Thus let $x \in N$, then we have

$$
\begin{aligned}
fh(x) &= \bar{f}h(x)\cdot q\bar{f}h(x) \\
&= \bar{f}h(x)\cdot q(\ell(x)\cdot y^{-1}) \quad \text{by (7.9)}, \\
&= \bar{f}h(x)\cdot q\ell(x) , \quad \text{since } y \in M_c , \\
&= \bar{f}h(x)\cdot \bar{q}(x) , \quad \text{by the definition of } q , \\
&= \ell(x) , \quad \text{by (7.9)}
\end{aligned}
$$

and the proof is complete.

<u>COROLLARY 7.3</u>. (Baer [7]) <u>Let</u> G/Z_kG <u>be finite of order</u> m <u>where</u> m <u>is a</u> P'-<u>number. Then</u> G^o_{k+1} <u>is finite of</u> P'-<u>order</u>.

<u>PROOF</u>: We proceed by induction on k. For $k = 0$, the assertion is obvious. Let $k \geqslant 1$. Considering G/ZG we know by induction that $G^o_k/G^o_k \cap ZG$ is finite. Since G/Z^o_kG is finite and $[G^o_k, Z^o_kG] = e$ it follows that G^o_{k+1} is finitely generated. It is thus enough to consider groups G that are finitely generated.

Set $N = Z^o_kG$ and $G/Z^o_kG = Q$. Denote the order of Q by m. Consider the extension $N \rightarrowtail G \twoheadrightarrow Q$. Theorem 7.2 yields a homomorphism $f : G \to N_P$. We may thus construct the diagram

$$(7.10)\qquad \begin{array}{ccccc} N & \overset{h}{\rightarrowtail} & G & \overset{g}{\twoheadrightarrow} & Q \\ \ell\downarrow & & \bar{f}\downarrow & & \| \\ N_P & \rightarrowtail & N_P \times Q & \twoheadrightarrow & Q \end{array}$$

with $\bar{f}$ the map given by f and g. Note that $\ker \bar{f} = \ker \ell$ is just the P'-torsion subgroup if N. But N, being a subgroup of finite index in a finitely generated group, is finitely generated. Hence $\ker \bar{f}$ is finite of P'-order. Since N_P is nilpotent of class at most k we have $(N_P)^o_{k+1} = e$, so that G^o_{k+1} is an extension of Q^o_{k+1} by a subgroup of $\ker f'$. It follows that G^o_{k+1} is finite of P'-order.

BIBLIOGRAPHY

[1] André, M.: Méthode simpliciale en algèbre homologique et algèbre commutative. Lecture Notes in Mathematics, Vol.32. Springer 1967.

[2] Atiyah, M.F., Macdonald, I.G.: Introduction to commutative algebra. Addison-Wesley 1969.

[3] Babakhanian, A.: Cohomological methods in group theory. Marcel Dekker Inc. 1972.

[4] Bachmann, F.: Kategorische Homologietheorie und Spektralsequenzen. Battelle Institute, Mathematics Report No.17, 1969.

[5] Baer, R.: Erweiterung von Gruppen und ihren Isomorphismen. Math. Z. 38 (1934), 375-416.

[6] Baer, R.: Representations of groups as quotient groups. I,II,III. Trans. Amer. Math. Soc. 58 (1945), 295-347, 348-389, 390-419.

[7] Baer, R.: Endlichkeitskriterien für Kommutatorgruppen. Math. Ann. 124 (1952), 161-177.

[8] Barr, M.: Exact categories. Lecture Notes in Mathematics, Vol.236. Springer 1971.

[9] Barr, M., Beck, J.: Seminar on triples and categorical homology theory. Lecture Notes in Mathematics, Vol.80. Springer 1969.

[10] Baumslag, G.: Some aspects of groups with unique roots. Acta Math. 104 (1960), 217-303.

[11] Baumslag, G.: Some remarks on nilpotent groups with roots. Proc. Amer. Math. Soc. 12 (1961), 262-267.

[12] Baumslag, G.: Some subgroup theorems. Trans. Amer. Math. Soc. 108 (1963), 516-525.

[13] Baumslag, G.: Groups with the same lower central sequence as a relatively free group. I. The groups. Trans. Amer. Math. Soc. 129 (1967), 308-321.

[14] Baumslag, G.: Groups with the same lower central sequence as a relatively free group. II. Properties. Trans. Amer. Math. Soc. 142 (1969), 507-538.

[15] Baumslag, G.: Lecture notes on nilpotent groups. Regional conference series in Mathematics, Vol.2. Amer. Math. Soc. 1971.

[16] Beck, J.: Triples, algebras and cohomology. Dissertation, Columbia, 1967.

[17] Beyl, F.R.: The classification of metacyclic p-groups, and other applications of homological algebra to group theory. Dissertation, Cornell, 1972.

[18] Chen, K.T.: Commutator calculus and link invariants. Proc. Amer. Math. Soc. 3 (1952), 44-55.

[19] Crowell, R., Fox, R.: Introduction to knot theory. Ginn and Company 1963.

[20] Eckmann, B.: Der Cohomologie-Ring einer beliebigen Gruppe. Comment. Math. Helv. 18 (1945-46), 232-282.

[21] Eckmann, B., Hilton, P.J.: On central group extensions and homology. Comment. Math. Helv. 46 (1971), 345-355.

[22] Eckmann, B., Hilton, P.J., and Stammbach, U.: On the homology theory of central group extensions: I. The commutator map and stem extensions. Comment. Math. Helv. 47 (1972), 102-122.

[23] Eckmann, B., Hilton, P.J., and Stammbach, U.: On the homology theory of central group extensions: II. The exact sequence in the general case. Comment. Math. Helv. 47 (1972), 171-178.

[24] Eckmann, B., Hilton, P.J., and Stammbach, U.: On the Schur multiplicator of a central quotient of a direct product of groups. J. Pure Appl. Algebra 3 (1973), 73-82.

[25] Eilenberg, S., MacLane, S.: Relations between homology and homotopy groups of spaces. Ann. Math. 46 (1945), 480-509.

[26] Eilenberg, S., MacLane, S.: Cohomology theory in abstract groups I, II. Ann. Math. 48 (1947), 51-78, 326-341.

[27] Epstein, D.B.A.: Finite presentations of groups and 3-manifolds. Quart. J. Math. Oxford (2), 12 (1961), 205-212.

[28] Evens, L.: Terminal p-groups. Ill. J. Math. 12 (1968), 682-699.

[29] Fox, R.M.: Free differential calculus, I. Ann. Math. 57 (1953), 547-560.

[30] Ganea, T.: Homologie et extensions centrales de groupes. C.R. Acad. Sc. Paris 266 (1968), 556-558.

[31] Gaschütz, W., Neubüser, J., Yen, T.: Ueber den Multiplikator von p-Gruppen. Math. Z. 100 (1967), 93-96.

[32] Gerstenhaber, M.: A uniform cohomology theory for algebras. Proc. Nat. Acad. Sci. U.S.A. 51 (1964), 626-629.

[33] Green, J.A.: On the number of automorphisms of a finite group. Proc. Royal Soc. London, Ser.A. 237 (1956), 574-581.

[34] Gruenberg, K.W.: Residual properties of infinite soluble groups. Proc. London Math. Soc. 7 (1957), 29-62.

[35] Gruenberg, K.W.: Cohomological topics in group theory. Lecture Notes in Mathematics, Vol.143, Springer 1970.

[36] Gut, A.: Zur Homologietheorie der zentralen Erweiterungen von Gruppen und von Lie-Algebren. Dissertation, Eidg. Techn. Hochschule, Zürich, 1973.

[37] Hall, P.: The classification of prime-power groups. J. Reine Angew. Math. 182 (1940), 130-141.

[38] Hall, P.: The splitting properties of relatively free groups. Proc. London Math. Soc. (3) 4 (1954), 343-356.

[39] Hall, P.: Nilpotent groups. Canad. Math. Congress, University of Alberta, 1957.

[40] Hall, P.: Some sufficient conditions for a group to be nilpotent. Ill. J. Math. 2 (1958), 787-801.

[41] Hilton, P.J.: Localization and cohomology of nilpotent groups. To appear in Math. Z.

[42] Hilton, P.J., Mislin, G., and Roitberg, J.: Homotopical localization. To appear in Proc. London Math. Soc.

[43] Hilton, P.J., Stammbach, U.: A course in homological algebra. Graduate Texts in Mathematics, Vol.4. Springer 1971.

[44] Hilton, P.J., Stammbach, U.: Two remarks on the homology of group extensions. To appear in J. Austr. Math. Soc.

[45] Hopf, H.: Fundamentalgruppe und zweite Bettische Gruppe. Comment. Math. Helv. 14 (1941/42), 257-309.

[46] Hopf, H.: Ueber die Bettischen Gruppen, die zu einer beliebigen Gruppe gehören. Comment. Math. Helv. 17 (1944/45), 39-79.

[47] Iwahori, N., Matsumoto, M.: Several remarks on projective representations of finite groups. J. Fac. Sc. Univ. Tokyo, 10 (1963/64), 129-146.

[48] Johnson, K.: Varietal generalizations of Schur multipliers, stem extensions and stem covers. To appear.

[49] Kervaire, M.A.: Multiplicateurs de Schur et K-théorie. Essays on Topology and Related Topics, Springer 1970, 212-225.

[50] Knopfmacher, J.: Extensions in varieties of groups and algebras, Acta Math. 115 (1966), 17-50.

[51] Knopfmacher, J.: Homology and presentations of algebras. Proc. Amer. Math. Soc. 17 (1966), 1424-1428.

[52] Lazard, M.: Sur les groupes nilpotents et les anneaux de Lie. Ann. Sci. Ecole Norm. Sup. 71 (1954), 101-190.

[53] Leedham-Green, C.: Homology in varieties of groups I. Trans. Amer. Math. Soc. 162 (1971), 1-14.

[54] Leedham-Green, C.: Homology in varieties of groups II. Trans. Amer. Math. Soc. 162 (1971), 15-26.

[55] Leedham-Green, C.: Homology in varieties of groups III. Trans. Amer. Math. Soc. 162 (1971), 27-34.

[56] Leedham-Green, C., Hurley, T.C.: Homology in varieties of groups IV. Trans. Amer. Math. Soc. 170 (1972), 293-303.

[57] MacLane, S.: Homology. Springer 1964.

[58] Magnus, W.: Ueber diskontinuierliche Gruppen mit einer definierenden Relation (Der Freiheitssatz). J. Reine Angew. Math. 163 (1930), 411-465.

[59] Magnus, W.: Beziehungen zwischen Gruppen und Idealen in einem speziellen Ring. Math. Ann. 111 (1935), 259-280.

[60] Magnus, W.: Ueber freie Faktorgruppen und freie Untergruppen gegebener Gruppen. Monatshefte für Math. und Phys. 47 (1939), 307-313.

[61] Mal'cev, A.I.: Nilpotent torsion free groups. Izv. Akad. Nauk. SSSR Ser. Mat. 13 (1949), 201-212.

[62] Meier, D.: Stamm Erweiterungen zur Primzahl p. Diplomarbeit, Eidg. Techn. Hochschule, Zürich 1973.

[63] Mostowski, A.W.: Automorphisms of relatively free groups. Fund. Math. 50 (1962), 403-411.

[64] Neumann, H.: Varieties of groups. Springer 1967.

[65] Neumann, P.: On word subgroups of free groups. Arch. Math. 16 (1965), 6-21.

[66] Neumann, P., Wiegold, J.: Schreier varieties of groups. Math. Z. 85 (1964), 392-400.

[67] Nomura, Y.: The Whitney join and its dual. Osaka J. Math. 7 (1970), 353-373.

[68] Rinehart, G.S.: Satellites and cohomology. J. Algebra 12 (1969), 295-329.

[69] Rinehart, G.S.: Oral communication, 1971.

[70] Robinson, D.J.S.: A property of the lower central series of a group. Math. Z. 107 (1968), 225-231.

[71] Schreier, O.: Ueber die Erweiterungen von Gruppen. I. Monatsh. Math. und Phys. 34 (1926), 165-180; II. Abh. Math. Sem. Hamburg 4 (1926), 321-346.

[72] Schur, J.: Ueber die Darstellung der endlichen Gruppen durch gebrochene lineare Substitutionen. J. Reine Angew. Math. 127 (1904), 20-50.

[73] Schur, J.: Untersuchungen über die Darstellung der endlichen Gruppen durch gebrochene lineare Substitutionen. J. Reine Angew. Math. 132 (1907), 85-137.

[74] Stallings, J.: Homology and Central Series of Groups. J. of Algebra 2 (1965), 170-181.

[75] Stammbach, U.: Anwendungen der Homologietheorie der Gruppen auf Zentralreihen und auf Invarianten von Präsentierungen. Math. Z. 94 (1966), 157-177.

[76] Stammbach, U.: Ein neuer Beweis eines Satzes von Magnus. Proc. Cambridge Phil. Soc. 69 (1967), 929-930.

[77] Stammbach, U.: Ueber freie Untergruppen gegebener Gruppen. Comment. Math. Helv. 43 (1968), 132-136.

[78] Stammbach, U.: Homological methods in group varieties. Comment. Math. Helv. 45 (1970), 287-298.

[79] Stammbach, U.: Varietal homology and parafree groups. Math. Z. 128 (1972), 153-167.

[80] Stammbach, U.: On the homology of a central subgroup. To appear.

[81] Swan, R.G.: Minimal resolutions for finite groups. Topology 4 (1964), 193-208.

[82] Tate, J.: Nilpotent quotient groups. Topology 3 (1964), 109-111.

[83] Ulmer, F.: Kan extensions, cotriples and André (co)homology. Lecture Notes in Mathematics, Vol.92. Springer 1969.

[84] Vermani, L.R.: An exact sequence and a theorem of Gaschütz, Neubüser and Yen on the multiplicator. J. London Math. Soc. 1 (1969), 95-100.

[85] Vermani, L.R.: An exact sequence attached to a nilpotent group. J. Fac. Sc. Univ. Tokyo 18 (1971), 329-333.

[86] Warfield, R.B.: Localization of nilpotent groups. Mimeographed, Seattle 1972.

Vol. 215: P. Antonelli, D. Burghelea and P. J. Kahn, The Concordance-Homotopy Groups of Geometric Automorphism Groups. X, 140 pages. 1971. DM 16,-

Vol. 216: H. Maaß, Siegel's Modular Forms and Dirichlet Series. VII, 328 pages. 1971. DM 20,-

Vol. 217: T. J. Jech, Lectures in Set Theory with Particular Emphasis on the Method of Forcing. V, 137 pages. 1971. DM 16,-

Vol. 218: C. P. Schnorr, Zufälligkeit und Wahrscheinlichkeit. IV, 212 Seiten. 1971. DM 20,-

Vol. 219: N. L. Alling and N. Greenleaf, Foundations of the Theory of Klein Surfaces. IX, 117 pages. 1971. DM 16,-

Vol. 220: W. A. Coppel, Disconjugacy. V, 148 pages. 1971. DM 16,-

Vol. 221: P. Gabriel und F. Ulmer, Lokal präsentierbare Kategorien. V, 200 Seiten. 1971. DM 18,-

Vol. 222: C. Meghea, Compactification des Espaces Harmoniques. III, 108 pages. 1971. DM 16,-

Vol. 223: U. Felgner, Models of ZF-Set Theory. VI, 173 pages. 1971. DM 16,-

Vol. 224: Revètements Etales et Groupe Fondamental. (SGA 1). Dirigé par A. Grothendieck XXII, 447 pages. 1971. DM 30,-

Vol. 225: Théorie des Intersections et Théorème de Riemann-Roch. (SGA 6). Dirigé par P. Berthelot, A. Grothendieck et L. Illusie. XII, 700 pages. 1971. DM 40,-

Vol. 226: Seminar on Potential Theory, II. Edited by H. Bauer. IV, 170 pages. 1971. DM 18,-

Vol. 227: H. L. Montgomery, Topics in Multiplicative Number Theory. IX, 178 pages. 1971. DM 18,-

Vol. 228: Conference on Applications of Numerical Analysis. Edited by J. Ll. Morris. X, 358 pages. 1971. DM 26,-

Vol. 229: J. Väisälä, Lectures on n-Dimensional Quasiconformal Mappings. XIV, 144 pages. 1971. DM 16,-

Vol. 230: L. Waelbroeck, Topological Vector Spaces and Algebras. VII, 158 pages. 1971. DM 16,-

Vol. 231: H. Reiter, L^1-Algebras and Segal Algebras. XI, 113 pages. 1971. DM 16,-

Vol. 232: T. H. Ganelius, Tauberian Remainder Theorems. VI, 75 pages. 1971. DM 16,-

Vol. 233: C. P. Tsokos and W. J. Padgett. Random Integral Equations with Applications to stochastic Systems. VII, 174 pages. 1971. DM 18,-

Vol. 234: A. Andreotti and W. Stoll. Analytic and Algebraic Dependence of Meromorphic Functions. III, 390 pages. 1971. DM 26,-

Vol. 235: Global Differentiable Dynamics. Edited by O. Hájek, A. J. Lohwater, and R. McCann. X, 140 pages. 1971. DM 16,-

Vol. 236: M. Barr, P. A. Grillet, and D. H. van Osdol. Exact Categories and Categories of Sheaves. VII, 239 pages. 1971. DM 20,-

Vol. 237: B. Stenström, Rings and Modules of Quotients. VII, 136 pages. 1971. DM 16,-

Vol. 238: Der kanonische Modul eines Cohen-Macaulay-Rings. Herausgegeben von Jürgen Herzog und Ernst Kunz. VI, 103 Seiten. 1971. DM 16,-

Vol. 239: L. Illusie, Complexe Cotangent et Déformations I. XV, 355 pages. 1971. DM 26,-

Vol. 240: A. Kerber, Representations of Permutation Groups I. VII, 192 pages. 1971. DM 18,-

Vol. 241: S. Kaneyuki, Homogeneous Bounded Domains and Siegel Domains. V, 89 pages. 1971. DM 16,-

Vol. 242: R. R. Coifman et G. Weiss, Analyse Harmonique Non-Commutative sur Certains Espaces. V, 160 pages. 1971. DM 16,-

Vol. 243: Japan-United States Seminar on Ordinary Differential and Functional Equations. Edited by M. Urabe. VIII, 332 pages. 1971. DM 26,-

Vol. 244: Séminaire Bourbaki - vol. 1970/71. Exposés 382-399. IV, 356 pages. 1971. DM 26,-

Vol. 245: D. E. Cohen, Groups of Cohomological Dimension One. V, 99 pages. 1972. DM 16,-

Vol. 246: Lectures on Rings and Modules. Tulane University Ring and Operator Theory Year, 1970-1971. Volume I. X, 661 pages. 1972. DM 40,-

Vol. 247: Lectures on Operator Algebras. Tulane University Ring and Operator Theory Year, 1970-1971. Volume II. XI, 786 pages. 1972. DM 40,-

Vol. 248: Lectures on the Applications of Sheaves to Ring Theory. Tulane University Ring and Operator Theory Year, 1970-1971. Volume III. VIII, 315 pages. 1971. DM 26,-

Vol. 249: Symposium on Algebraic Topology. Edited by P. J. Hilton. VII, 111 pages. 1971. DM 16,-

Vol. 250: B. Jónsson, Topics in Universal Algebra. VI, 220 pages. 1972. DM 20,-

Vol. 251: The Theory of Arithmetic Functions. Edited by A. A. Gioia and D. L. Goldsmith VI, 287 pages. 1972. DM 24,-

Vol. 252: D. A. Stone, Stratified Polyhedra. IX, 193 pages. 1972. DM 18,-

Vol. 253: V. Komkov, Optimal Control Theory for the Damping of Vibrations of Simple Elastic Systems. V, 240 pages. 1972. DM 20,-

Vol. 254: C. U. Jensen, Les Foncteurs Dérivés de $\varprojlim$ et leurs Applications en Théorie des Modules. V, 103 pages. 1972. DM 16,-

Vol. 255: Conference in Mathematical Logic - London '70. Edited by W. Hodges. VIII, 351 pages. 1972. DM 26,-

Vol. 256: C. A. Berenstein and M. A. Dostal, Analytically Uniform Spaces and their Applications to Convolution Equations. VII, 130 pages. 1972. DM 16,-

Vol. 257: R. B. Holmes, A Course on Optimization and Best Approximation. VIII, 233 pages. 1972. DM 20,-

Vol. 258: Séminaire de Probabilités VI. Edited by P. A. Meyer. VI, 253 pages. 1972. DM 22,-

Vol. 259: N. Moulis, Structures de Fredholm sur les Variétés Hilbertiennes. V, 123 pages. 1972. DM 16,-

Vol. 260: R. Godement and H. Jacquet, Zeta Functions of Simple Algebras. IX, 188 pages. 1972. DM 18,-

Vol. 261: A. Guichardet, Symmetric Hilbert Spaces and Related Topics. V, 197 pages. 1972. DM 18,-

Vol. 262: H. G. Zimmer, Computational Problems, Methods, and Results in Algebraic Number Theory. V, 103 pages. 1972. DM 16,-

Vol. 263: T. Parthasarathy, Selection Theorems and their Applications. VII, 101 pages. 1972. DM 16,-

Vol. 264: W. Messing, The Crystals Associated to Barsotti-Tate Groups: With Applications to Abelian Schemes. III, 190 pages. 1972. DM 18,-

Vol. 265: N. Saavedra Rivano, Catégories Tannakiennes. II, 418 pages. 1972. DM 26,-

Vol. 266: Conference on Harmonic Analysis. Edited by D. Gulick and R. L. Lipsman. VI, 323 pages. 1972. DM 24,-

Vol. 267: Numerische Lösung nichtlinearer partieller Differential- und Integro-Differentialgleichungen. Herausgegeben von R. Ansorge und W. Törnig, VI, 339 Seiten. 1972. DM 26,-

Vol. 268: C. G. Simader, On Dirichlet's Boundary Value Problem. IV, 238 pages. 1972. DM 20,-

Vol. 269: Théorie des Topos et Cohomologie Etale des Schémas. (SGA 4). Dirigé par M. Artin, A. Grothendieck et J. L. Verdier. XIX, 525 pages. 1972. DM 50,-

Vol. 270: Théorie des Topos et Cohomologie Etale des Schémas. Tome 2. (SGA 4). Dirigé par M. Artin, A. Grothendieck et J. L. Verdier. V, 418 pages. 1972. DM 50,-

Vol. 271: J. P. May, The Geometry of Iterated Loop Spaces. IX, 175 pages. 1972. DM 18,-

Vol. 272: K. R. Parthasarathy and K. Schmidt, Positive Definite Kernels, Continuous Tensor Products, and Central Limit Theorems of Probability Theory. VI, 107 pages. 1972. DM 16,-

Vol. 273: U. Seip, Kompakt erzeugte Vektorräume und Analysis. IX, 119 Seiten. 1972. DM 16,-

Vol. 274: Toposes, Algebraic Geometry and Logic. Edited by. F. W. Lawvere. VI, 189 pages. 1972. DM 18,-

Vol. 275: Séminaire Pierre Lelong (Analyse) Année 1970-1971. VI, 181 pages. 1972. DM 18,-

Vol. 276: A. Borel, Représentations de Groupes Localement Compacts. V, 98 pages. 1972. DM 16,-

Vol. 277: Séminaire Banach. Edité par C. Houzel. VII, 229 pages. 1972. DM 20,-